READY FOR A CAREER CHANGE?

INTERVIEWS WITH SUCCESSFUL CAREER TRANSITIONERS, AND 9 LANDMARK QUESTIONS TO GET YOU THROUGH A CAREER CHANGE IN ONE PIECE

SARAH O'FLAHERTY

CONNECT THE DOTS

JOIN MY VIP CLUB

.

1

WHAT ARE WE SEARCHING FOR?

What is it you are seeking? You must be searching for something or you wouldn't be reading this book. Doesn't it often seem like there is something just outside your view that you need to find, but you're not quite sure what it is? You know your life should be different, better, than it is right now, but you're not quite sure what to change to find a higher level of satisfaction. And, do you know that if you don't try to find that little niggling sense of 'there must be something more', it will never go away? I know, because I tried to ignore that feeling for a long time, and it did not depart.

It might just be that random questions pop into your head every now and then. Questions like, why haven't I found contentment yet? Or, isn't life meant to be a bit more interesting and exciting than this? Why am I putting up with hour-long commutes, and this prison-style culture of nine-to-five office life? If you think and feel that you're missing out on something, then it's very likely that you are. But what is it? What is the enigmatic *something* that we're all craving?

I believe it's related to our development or evolution as a human being. We are all unique and we all have our own specific reason for being on this wonderful planet. Unfortunately, for so many of us, we haven't yet found our purpose, our meaning. My concern, and the reason why I've written this book, is that there are far too many of us who aren't even trying to find it. What is stopping you? Fear, worry that change will make things worse not better, or just complacency? Unless we find our own unique purpose and the amazing value we must offer others, we may be nothing more than cogs in the industrial machine – keeping the dollars flowing in for others and not really doing much to make the world a better place.

Take a moment to think about what you do for work. Do you consider it 'right' work? Is it making a positive impact on the world, and are you making the world a better place, or are you adding to the negativity and environmental damage that is growing around us? You may believe that what you do doesn't make any difference in a world that's so full of people, but what each and every one of us does makes a huge difference – you included. If you haven't taken a moment to think about what you do for work, then do it now. It's worth a few moments of contemplation.

It's in these times of change – and things are changing rapidly right now – that we are provided with the greatest opportunities to find our own true bliss, our own unique and original purpose. There are many examples of people who are doing just that – finding their own unique reason for being in the world and building their dream life.

I set out to find some of these amazing individuals, to uncover their journeys of courage, and to share their stories so you know you can do it too. These are not celebrities, or the lucky few who have made it big on a large scale, these are ordinary folk who are making a positive difference in the world in their own way, and who are living their dream lives. Some examples of the stories included in this book are a publicist turned reiki teacher, a model turned youth role model and trainer, and an advertising exec turned coaching entrepreneur. I hope these examples, being more down-to-earth, will help you see that if they can do it, you can too.

My own journey has been an interesting adventure that I'll share briefly before we hear from some of the inspiring individuals I've mentioned above. I only share my story to give you some perspective on the changes I've been through to get to where I am now. I also think it's helpful to know that life doesn't always follow a clear and obvious path – it sometimes takes us down some very rocky roads before we get to a place of peace and contentment. And it's important to be open-minded about where we are heading, as it may be very different to what we expected to find when we set out on our journey.

Let's start from the beginning. I'm from a small town in New Zealand (NZ) and was brought up in an ordinary, middle-class family. I would say that I had a reasonably sheltered upbringing, living on a farm near a small rural town. I had a happy, stable childhood that was rather unexceptional. Both my parents went to university and I was expected to do the same. My father was a veterinarian and my mother a primary school teacher. There was initially an expectation that I would become a vet like my father. However, luckily, my lack of interest in that area and my inability to handle the sight of blood meant that particular option came off the table very quickly.

As a young girl I dreamed of becoming an international businesswoman – travelling the world, dressing in stylish suits, and earning lots of money. Where that idea came from I'll never know. Possibly from TV shows or magazines, because I didn't know what it meant to be an international businesswoman except that I thought it would mean lots of travel and getting to dress rather glamorously.

I've always had and continue to have a desire to travel and explore. Therefore, the travel part of my early dreaming was pretty spot on. The business part was also not bad for a very naïve guess – I've always had an interest in business and an entrepreneurial spirit. And even now, while studying, I've been running my own business and helping others to establish theirs. So although my early dreaming meant that I went to university and completed a business degree with a focus on international management, which I thoroughly enjoyed, I hadn't really got clear on what was important for me personally and what I wanted or needed in the longer term.

I started out working in marketing for a large corporation in NZ, and after only a couple of years shifted into the much more exciting world of advertising. I loved this new world I'd joined – I loved the energy, the creativity, the people – it was fun, fast-paced, and exciting. However, I hadn't thought about it too much before taking the leap. I'd just thought, this feels good, and started climbing my way up the corporate ladder. At the time, as with many young people, I would say that I wasn't totally clear on what my values were or what was important to me. I made my decision based on the challenge of the work, the wonderful environments I got to work in, and the intelligent, sparky people I was lucky enough to work with. All good reasons at the time, but I didn't account for other things. My generally introverted personality was bombarded with an assortment of human interactions all day, and then I was often required to entertain clients in the evening. And what about my values? At no time did I think, is this job doing good for the world or am I perhaps having a negative impact on people's lives?

Over time the industry changed, becoming more and more money orientated and less and less creatively driven. Eventually, it got to the point where everyone wanted wonderful work for zero dollars. I found the money versus output focus of the industry demoralizing and frustrating. Most of all, as I became increasingly anti-consumerism, selling products to people who didn't need them didn't sit at all well with my values. It took many difficult and stressful situations, finally culminating in a slipped disc in my neck, to force me to pull the pin on a career that had taken me all over the world and pretty high up the corporate food chain.

By that stage I'd known for quite some time that I wasn't following my bliss. You might well ask why I was still working in a career that clearly didn't suit me anymore. Well, let's be honest, it was partly the money. I was being paid very well to do a job that I knew inside-out, and at the time, I didn't know what else to do with myself. I wasn't one of the lucky ones who'd stumbled upon their bliss early in life – who'd found their purpose in their first career choice.

Being an avid Joseph Campbell fan, the man who coined the term *finding your bliss*, I'd always dreamed of finding my bliss, and eventually I knew it was time to go on my own 'hero's journey' to find out why I had been put on this earth. I talk more about the hero's journey shortly, but for now let me just say, as with the hero's journey, once I'd decided to follow the call to adventure I was tempted by offers that could easily have distracted me from my mission to find my true purpose and more energised way of living.

I resigned from my job, by this stage I was working in Thailand in a regional role, after deciding to take a year off to travel. It was then that I was offered my dream job – a chance to go to Peru to work for six months to a year. Wow, how amazing. Peru was the one place in the world I'd been dying to visit – the idea of trekking up to Machu Picchu was a dream come true. As you can imagine, it was very tempting and I nearly got sucked back into the void.

But then I stopped and thought about it. I knew that I no longer wanted to work in advertising, and I knew that even the enticement of what appeared to be the best job in the world wasn't worth it. If I really thought about it, I knew that I could go to Peru any time from anywhere in the world and be there on my terms. I didn't have to take up the offer to go there in a company role - so I declined. I packed up my life in Thailand, shipped it home to New Zealand, and went off to Bali for the start of my year of adventure travelling around Asia.

The next year was wonderful. It was one of the best years of my life, and I wondered frequently during that time why I hadn't taken a year off from work earlier. It was that year of travel that allowed me to find my true bliss.

"If you do follow your bliss you put yourself on a kind of track that has been there all the while, waiting for you, and the life that you ought to be living is the one you are living. Follow your bliss and don't be afraid, and doors will open where you didn't know they were going to be."

Joseph Campbell

The Hero's Journey

Joseph Campbell is one of my heroes. His writing is vast and covers a wide range of human experiences. If you get a chance, I highly recommend taking the time to read one of his many books. He is the creator of the term 'finding your bliss', and he established the famous 'hero's journey'. In case you aren't aware of what the hero's journey is, let me provide you with a brief overview. And, please note, where I write 'hero' I am talking about a male or female hero. 'Heroine' seems like such an old-fashioned word to me, and using hero keeps things simple. I like simple. I'm going to use 'her' for the hero's journey segment, but feel free to replace that term with whatever is appropriate for your gender.

Campbell studied mythology and symbology for many years. His interest began when he spent about four years reading in a cabin in the woods during the Great Depression. It was during that time that he identified some common themes running through the hero myths and legends from around the world. He identified a set of stages that a hero must go through on his journey, on his own unique quest. Interestingly, these stages are the same no matter in what country the story originates.

Campbell's book *The Hero's Journey* has been the inspiration for many famous writers and film-makers. George Lucas, the creator of Star Wars, said that *The Hero's Journey* was the inspiration for his famous movies. It can be speculated that it is because George Lucas used such a time-honored structure that his films worked so well and became so iconic and globally popular. As we walk through each of these stages, see if you can identify how they relate to you.

The journey begins in what is considered the **ordinary world**, the world we currently operate in, the generally uneventful world of routine day-to-day living. The hero may be considered a bit unusual or odd by those operating within the norms of society. Additionally, the hero often possesses some ability or characteristic that may make her special and different, but may also make her feel out of place. A good example is Dorothy in *The Wizard of Oz*, in Kansas before the cyclone strikes – she is alone, perhaps a bit isolated, and feeling out of place in her world, where she doesn't seem to quite fit in.

The next stage is when our hero receives a **call to adventure**, and is called away from the ordinary world to begin their quest. The hero may initially show some reluctance to leave their home, friends, and family to head away, but usually they will accept the quest that is their destiny. The hero may happen upon their quest by accident, or may be called to the quest to save their world.

The quest takes place in another world, which Campbell describes as a "fateful region of both treasure and danger...a distant land, a forest, a kingdom underground, beneath the waves, or above the sky, a secret island, lofty mountaintop, or profound dream state...a place of strangely fluid and polymorphous beings, unimaginable torments, superhuman deeds, and impossible delight." This description appears to cover many magical realms. It reminds me of the magical world Alice discovers in the story of *Alice in Wonderland*.

Next there is the **refusal of the quest** or the **challenge to the quest**. This stage occurs when the hero is called to adventure and given a task or quest that only they can complete. At this stage, they do have a choice – they can accept the quest or deny it. Although this may seem like a simple matter when looked at superficially, it is not as clear-cut as you might imagine. The hero may be tempted by another offer, or possibly a reward to stay at home, or they may decide that they don't want to accept their destiny. Remember the temptation I had to go to Peru rather than heed my call to adventure.

Unfortunately, for those heroes who decide that they do not, for whatever reason, want to accept their destiny, the future is not a rosy one. These individuals generally end up being the characters in need of rescuing or may even end up as the villain in a future tale.

Let me interrupt the story for just a moment, and ask you, does this remind you of anyone? Do you know of people who have been offered an amazing opportunity that required considerable life changes but also offered huge opportunity for growth, who were not brave enough to take the opportunity when it was presented, and ended up becoming bitter and frustrated? For those accepting the call, sometimes it is a matter of the call having to be presented a number of times until it is finally accepted.

As the hero embarks on her journey she enters the **world of the unknown**, a world that may be filled with supernatural creatures, spectacular vistas, adventures, and dangers. This new world will have a set of rules that are different from the hero's home world. The hero learns these rules as she progresses on the journey. This is the start of her learning process, which continues through the entire quest experience.

It is at this 'point of entry' into the new world that the hero is likely to **meet their mentor.** This stage has also been interpreted to mean that the hero will receive some form of supernatural assistance before beginning their quest. The mentor has already mastered the new world and can provide the hero with confidence, insights, advice, training, and sometimes magical gifts that may overcome future challenges. The mentor shares his experience and knowledge so the hero is not rushing blindly into the new world. The mentor often provides a gift or some form of wisdom that is required for the quest to be completed. To quote Campbell himself, "One has only to know and trust, and the ageless guardians will appear." A good example of the mentor archetype is Gandalf from *Lord of the Rings.*

The next step is **crossing the threshold,** where the hero commits to the journey. The hero is now ready to cross the gateway that separates the ordinary world from the world of the quest. The crossing may require the acceptance of one's fears, a map of the journey, or perhaps an incentive offered by someone else. The hero must confront an event that forces her to commit to entering the new world. At this point, there is no turning back. The event will directly affect the hero, raising the stakes and forcing some action. The event may be an outside force, such as the abduction of someone close to the hero that pushes her ahead. Or there may be a chase that pushes the hero to the brink, forcing her to move forward and commit fully to the quest.

After crossing the threshold, the hero faces **tests,** encounters **allies,** confronts **enemies,** and learns the rules of the special new world she has entered. This is the introduction to the new world, and we can see how it contrasts with the ordinary world the hero has come from. It is at this stage in the journey that the hero determines who can be trusted. A sidekick may be found, or even a full hero team developed. Enemies reveal themselves, and a rival to the hero's goal may emerge. The hero must begin to prepare herself for the greater ordeals to come, and so tests her skills, and, if possible, receives further training from the mentor.

The hero must then **approach the innermost cave** that leads to the journey's heart or the central ordeal of the quest. Maps are reviewed, attacks planned, and the enemy's forces whittled down before the hero can face her greatest fear or the supreme danger that is lurking. The approach offers a chance of a break for the hero and her team before the final ordeal. The team may need to regroup, remember the dead, and rekindle morale. At this point in the journey time may be running out, or the stakes may rise.

The hero reaches **the ordeal**, the central life-or-death crisis during which she faces her greatest fear and confronts her most difficult challenge. She may experience a form of 'death'. She may teeter on the brink of failure, and we may wonder if our hero will survive. It is only through 'death' that the hero can be reborn, experiencing a resurrection that offers greater powers or insight to see the journey through to the end. This is often the ultimate battle of good versus evil.

Once the hero has survived death and overcome her greatest fear or fears she will earn **the reward** she has been seeking. The reward may come in many forms – it may be a magical weapon, a secret potion, knowledge or wisdom, or the return of relationships lost. No matter what the treasure, the hero has earned the right to celebrate. The celebration gives the hero a new burst of energy.

Next, the hero must recommit to finishing the journey and taking **the road back** to the ordinary world. The hero's success in the other world may make it difficult for the hero to return home. Like crossing the threshold, the return home may require a special event or something that pushes the hero back toward home. This event may re-establish the central dramatic question, pushing the hero to action and raising the stakes. As with any strong turning point, the action that galvanizes the road back may very well change the direction of the story.

The hero then faces **the resurrection**, another dangerous meeting with death. This is the final life-and-death ordeal that shows that the hero has learnt and maintained what she needs to bring back with her to the ordinary world. This is a final 'cleansing' or 'purification' that must occur now that the hero has emerged from the other world. The hero is reborn or transformed with the addition of the lessons and insights learned from her journey.

Once the hero has been resurrected, she has earned the right to **return with the elixir** to the ordinary world. The elixir can be a great treasure or magic potion, it could be love, wisdom, or simply the experience of surviving the other world. The hero may share the benefit of the elixir, using it to heal a physical or emotional wound, or to accomplish tasks that had previously been considered impossible in the ordinary world. The return signals a time when rewards and punishments are dished out, and the end of the journey is celebrated. The return with the elixir generally brings closure to the story and balance to the world. The hero can embark on a new life knowing she has survived the trials of her journey.

"All the world's a stage. And all the men and women merely players: They have their exits and entrances; And one man in his time plays many parts."
William Shakespeare (As You Like It)

Throughout the journey there are many archetypes or roles that characters play. You can consider each archetype as a mask that a character wears in a scene. Sometimes the character may wear the same mask throughout the story. However, as with life, each character may play many roles throughout the journey and represent different archetypes. The key archetypes are listed below:

1. Hero – to serve and sacrifice
2. Mentor – to guide
3. Threshold Guardian – to test
4. Herald – to warn and challenge
5. Shapeshifter – to question and deceive
6. Shadow – to destroy
7. Trickster – to disrupt
8. Allies – to offer support

You may find that you face similar archetypes in your own journey through life.

What I love about the hero's journey is that it is surprisingly reflective of what happens in our own lives. You may think, no way, I've never journeyed to another world, I've never fought dragons, I've never died and been resurrected. But think of the hero's journey as a metaphor for life, with each stage of the journey symbolic of a stage of life. It does make sense. I can certainly follow my own life journey in parallel to the hero's journey. And, to be honest, I find the process of assessing my own life journey against the hero's journey to be very useful. It allows me to see the benefit of challenges I've faced, where I'm up to in any particular journey or life stage, and in what ways I'm meant to be evolving. It even gives me clarity about my life purpose. Here's the start of my most recent journey, as a brief example:

Ordinary world: Working in advertising.

Call to adventure: Slipped disc in my neck (the call to adventure can be related to poor health, a bad accident, or some form of near-death experience). For me it was the final sign that I needed to leave the world of advertising and move on.

Challenge to the quest: Following my resignation from work, my employers offered to send me to work in Peru, my dream destination. Luckily I was aware of the hero's journey and saw this as a challenge to my need to move on. Rather than go to Peru on work's terms, I decided I would go on my own when it suited me.

Entering the world of the unknown: Taking a year off to travel around Asia. At this point I had no idea what I was going to do next with my life.

Meeting the mentor: Meeting a Buddhist teacher who assisted me in my decision to stay in the world to help others rather than becoming a Buddhist nun (which I was considering).

Crossing the threshold: Deciding that I would definitely leave my career in advertising and explore my options.

Tests, allies, enemies: I have certainly faced some challenges on this journey. And I've met some people that I believe could be called enemies, including having a relationship with someone with borderline personality disorder. This person was incredibly dishonest and deceitful, and did his best to destroy me (yep, just like the bad guys in the movies). Luckily, I finally saw the dark side and escaped relatively unscathed. I also developed strong allies along the way, by building very strong friendships and support networks.

Approaching the innermost cave: Committing to six years of training to become a clinical psychologist.

The ordeal: The ordeal for me has been finding a new career path, choosing that path, and then sticking with it. Deciding to take six years away from working to study for a new profession was a huge decision, and of course the financial implications loomed large. It has been a challenging and uncomfortable process.

The reward: I can happily say that already my life is so much richer and more satisfying by having taken my journey to this point. The reward, I believe, is still to come. I am hoping that the reward will become clear when I have completed my training. I feel it will be the service I will be able to provide to others in helping them deal with whatever suffering is affecting them in their lives.

My journey continues, and I certainly believe that this path, while not the easiest one, has helped me evolve into a better human being. It's important to be aware that it's possible to go through more than one hero's journey in a lifetime. As I've mentioned, times are changing fast, and the energy of the world is pushing us to keep up. These changes are likely to mean more challenges and more personal evolution for us all.

If you take up the challenge to find your own unique way in the world you will reap the benefits. Ultimately, it's important to remember that life is one big adventure and we all get offered the opportunity to go on a hero's journey. While many of us choose to stay in the ordinary world, it is possible that we do this to our own detriment. For unless we take up the challenge and face our fears, we will not bring back the ultimate reward of becoming the best person we can be in this lifetime.

2

DISILLUSIONED

For many of us, finding our bliss, going out alone, following our passions, seems like a crazy fantasy. It's an impossibility that we shouldn't even imagine. We think we don't deserve it, or we think it's a dream that's only possible for the already rich, or perhaps the very lucky few. And yet, with so many of us disillusioned with working for corporations, institutions, and dysfunctional not-for-profits, you'd think more people would be braver and at the very least give trying out some other options a go.

Let's start by looking at what's involved in working for someone else. To work in today's world many of us need to live in large cities where employment is often located in the central business district. What this means for most people is getting up early and catching public transport or driving for an hour or more to get to work. A recent trip to visit a friend in the outer suburbs of Sydney reminded me of what that commuting experience can be like. We decided to catch a train to the central business district, and, sadly, we did this during rush hour – not a good decision. We were crammed into a packed train for nearly an hour. I can honestly say I felt like I'd been squeezed into a tin of sardines – except all those other sardines were human beings.

Phew, it was a tough ride and not a pleasant experience. And, to top it off, I was very aware that I only had to do it for one day, not five days a week for most of the year.

Then there's the nine-to-five routine. Who on earth decided that we must work from nine to five, five days a week? Many of us work more like eight to eight most days, with our work over-flowing into the weekend. I feel like humankind has developed a kind of institutionalized prison system and called it work. I'm not sure where this regimented system came from, but it doesn't feel healthy or normal or necessary.

If you deconstruct the nine-to-five cubicle culture, it appears to have been established to ensure we work like machines. Generally, there are walls surrounding you so you can't interact with anyone else – we don't want you wasting your time chatting! Each spot is similar, so you can't be differentiated from anyone else. And it is possible that you may end up sitting at a computer all day – which isn't healthy for anybody. These types of offices are generic, anti-social and uncreative. And yes, as you can tell, I'm quite anti-establishment. I think it's time to get the humanity back into the workplace and organizational cultures. And yes, I know there are some workplaces that manage to avoid this bland, homogenous environment – take Google as an example – but sadly, these companies are few and far between. When will we think a little differently? Operate more flexibly? Allow people to work from home sometimes?

Sadly, the money we earn in our job very often goes to pay for the fancy clothes we wear to work, the cost of travel to and from work, and the lunches we buy at work. Many of us have families but lack quality family time as work takes priority over every-thing else. Furthermore, due to the stress of many jobs, we also spend a lot of our earnings on doctors, medicines, therapy, massages, and counselling. If we're lucky we'll have some money left over for a holiday once a year.

I remember watching a documentary on happiness that

showed some of the happiest people in the world living in a small village in Italy. In that little village, each person had their own specific role to play. The documentary focused in part on an old lady who looked after the village goats. She would get up early and take them out to the fields to graze for a couple of hours (she got exercise and fresh air). She would then head home, and spend a few hours weaving the goats' wool (producing something useful). In the evening she hung out with the rest of the village (social interaction and community). Each person in the village only worked for about five hours a day, and yet they each had a roof over their head and good food on their table. They lived in a supportive community, and had a wonderful, simple lifestyle. When asked what they would do if they won the lottery, almost all the villagers said they wouldn't change a thing, that they loved the life they had. Can you say the same?

When you look at the happiest people in the world – the ones living the simplest lives – and compare them to those of us living in the so-called modern world of gadgets, celebrity, and high salaries, you must wonder who is winning at the game of life. Why have we moved away from this simple way of life? Is it satisfying to just keep on purchasing more and more stuff? Are we living consciously? Or are we being swept up by the fads of today that encourage us to believe we must have the best of everything, that we will never be satisfied unless we have more, more, and more. What kind of life are you living? Is it fun, is it satisfying? Really? Is it the life you thought you'd be living as you grew up?

This brings back yet another memory, of a conversation I had with colleagues over fifteen years ago when I was working London. At that time there was serious talk of the possibility of a paperless society, and things were changing so quickly (if we only knew then how slow the change was compared to what we are now experiencing) that it was thought we would soon be working fewer hours, allowing more people to be gainfully employed, with there would be more flexi-time. Work locations would also be more flexible – more people could work from home if they chose to. What happened to these discussions and the opportunity for increased work–life balance? It appeared that things were going to get better – that we would all be working fewer hours with more leisure time. When. But in reality, there are fewer people working, those people are working longer hours, more and more locations are operating like sweatshops rather than offering flexi-time and flexi-location work, and people have limited holiday options and a fairly restricted life.

I don't want to get you down with this conversation – this is not everyone's experience. This is the world of some corporate environments; the life of those who are not following their dreams. Luckily there are now other options emerging that allow us to work fewer hours, and in the location of our choice – like me, currently working from home to write this book. And that's why I've written it – so you can see how other people are living their passion and develop the skills to do it yourself.

3

MYTH BUSTING

Like everything, there are plenty of myths surrounding the idea of following your bliss or your passion. Most of these are made up to scare you, some have the tiniest hint of truth in them, and some of them seem to have emerged out of thin air. However, no matter where the myths come from, they need to be busted so that you can feel free to follow your dreams. The trouble with myths is they are easy to believe and can sometimes stop us before we make much progress. Let's get started and break down some of these mental barriers.

Myth #1: You Will Not Make Money

I think it's a good idea to bust the biggest myth first. This is the one most people are concerned about; if I follow my dream how can I possibly make enough money to live? Let me try to allay your fears. You can and will make enough money to live on if you follow your dreams – if you truly find your unique purpose and the unique value you offer the world. When you start doing what you love it drives you, you put more energy into it, and that energy pays off in growth and output. It may take a little time initially to get going – like any business that's just kicking off, your dream job or business needs time to sprout and grow before it can bear fruit.

Many people start moving into their dream job while still working in their current role. It's hard going at first while you're working on two things at the same time, but once you know you're doing the right thing, and it's starting to show returns, you can make the shift and suddenly you'll be living your dream life. I recently heard of a lady who decided she wanted to become a self-published author. She hired an expert to help her map out a plan to reach that goal. This expert recommended a genre and helped her with her marketing. She started writing while still working and within a year she was making a six-figure income, could quit her job, and is now living her dream life. Amazing, isn't it? Of course, these great results did not just fall into her lap. She was smart about it – she hired an expert to help her get started, and she worked damn hard that first year. You'll find in this book many examples of people following their bliss, living the life of their dreams, and making money doing it.

Now, it is important to remember that not everyone mentioned in this book is making a fortune. Some are just making enough to get by, but are happy with the life they're leading; they live simply. I think at this point it's good to question whether all your expenses or outgoings are necessary. Many of us are living way outside our means and that isn't helping us to get on track with our life purpose. When you slow down, simplify, and live the life of your dreams you may find you don't need even half the income you had in the past to survive. I'm not just making this up, I'm talking from experience. I have chosen to stop earning the big bucks from my career in advertising, to retrain to become a psychologist, and have been surviving as a student on only a small amount of part-time work. Of course, I have also used some of my savings to make this happen. But when you know you're on the right track, you also know it's worth it, and you're worth it. Over the past couple of years, I have survived on thirty per cent of what I was earning while working in advertising, and I have never been happier or more content.

Myth #2: You Must Know What You Want To Do....Now

Another myth that has infiltrated our thinking is the myth that you must know what you want to do, and you must know it right now. If you are unsure of what your bliss is, or you haven't found your purpose yet, don't worry, there is still hope. You may be one of the many who take a bit longer to find their bliss. It took me a very long time – after a twenty-year career in advertising, I finally took that year off work to travel and write, and that was when I found my bliss.

If you're still seeking your bliss, you're in good company – there are plenty of people still searching. Don't forget all the wonderful examples of people who succeeded later in life. Ray Kroc had passed his fiftieth birthday before he bought the first McDonald's. Sylvester Stallone was "broke" for a long time before he became a famous actor. At one stage, he was so poor that he had to sell his wife's jewellery and his dog to pay his bills. He became successful when the script for the movie Rocky, which he wrote, was accepted; he was 30 years old. Another great example is J. K. Rowling, the writer of the Harry Potter series. She struggled for years and was rejected many times before she finally managed to sell her first story. That sale happened in her mid-thirties, and she has since gone on to become a billionaire. After going from one failure to another, Mark Pincus, the creator of Zynga, finally benefitted from his passion for gaming, which led him to set up his own successful company when he was in his forties. And finally, one of my favorite later-in-life success stories is Anna Mary Robertson Moses, also known as Grandma Moses. She began her extensive painting career at seventy-eight years old. In 2006, one of her paintings sold for US$1.2 million.

There's always time to find and follow your passion. The key is to not give up, and keep seeking out your bliss. You have been put on this earth for a reason and it can take time to find it. There are plenty of ways to find your bliss, and one I strongly recommend is to wander, roam, explore – travel the world if you can, explore different careers, see what the people you envy are up to. What are they doing that you wish you were doing, what path did they travel to get to where they are today? And if you think you've found your passion already, but it's not yet returning the rewards you hoped it would, remember that you are doing something that you love, and focus on how that makes you feel. Doing what you love gives you energy, while doing work you hate will suck the life out of you.

You will find I mention Joseph Campbell a lot in this book, he is the man who coined the term *follow your bliss*. I like this quote from him in relation to wandering:

When you wander, think of what you want to do that day, not what you told yourself you were going to want to do. And there are two things you must not worry about when you have no responsibilities: one is being hungry, and the other is what people will think of you. Wandering time is positive. Don't think of new things, don't think of achievement, don't think of anything of the kind. Just think, "Where do I feel good? What is giving me joy?"

I do recommend reading some of Joseph Campbell's books if you can – they are packed full of great wisdom. A couple that you might enjoy are *Myths to Live By* and *A Joseph Campbell Collection: Reflections on the Art of Living*. Also, don't forget to check out my first book, *Simplify Your Life,* available at all ebook retailers or for free at my website, for a comprehensive process that will help you lead a more satisfying life. I am also developing an online course that expands on the book, with more tools, exercises, and added-value material. Information for the course can be found on my website at sarahoflaherty.com.

Myth #3: You Must Be Creative

Creativity and creative thinking are becoming more and more desirable in a world where basic tasks are being automated and innovation is considered king. And yet, for some reason, many of us seem to have a real block around the words *creative* and *creativity*. It seems we've been trained by society and our school systems to think that we aren't and can't be creative. Here's the myth-busting – in one way or another, we are all creative. I know, it's easy to say, but it's true. Every single one of us is creative in our own unique way. And it's important that you believe you are creative – because if you don't believe you are, then you won't be.

It's possible that we're stuck on the term 'creativity' because we think it means that we must be an amazing artist, musician, or actor. But creativity should be considered a phenomenon whereby something new and somehow valuable is formed – this can be an idea, a design, a piece of art, a new method of tackling a task, absolutely anything at all. I am a huge fan of Sir Ken Robinson, who is an author, speaker, and international advisor, and a strong advocate of creativity. He says that "one of the enemies of creativity and innovation, especially in relation to our own development, is common sense." We get stuck in our logical mind, and don't accept our intuition and our creative urges. I believe that another enemy of creativity is our concern about what others might think of us. We often hold ourselves back for fear of being judged or criticized.

Remember, to follow your bliss you don't have to be a "creative", as in an artist, a musician, or an actor. We are all creative, and tapping into that creativity can sometimes open us up to our purpose.

Myth #4: Thou Shalt

Another myth, or what I consider to be a major block for people, is the 'thou shalt' barrier. Thou shalt is the weight of society bearing down on us, telling us how we should be and what we should do. There is a wonderful fable that beautifully illustrates this myth. The story begins with a baby camel born in the desert. As a baby the camel is free to wander, roam, and explore his little part of the desert. Life is easy and simple. As the camel grows older he starts to be used by humans for carrying baggage – every day, a small amount of baggage is loaded on his back. (Think of this baggage as the expectations and mandates of society.) As the camel gets older and bigger, more and more baggage is loaded on his back, until eventually he's carrying all he can bear. One day the camel is wandering through the desert loaded high with baggage, and he notices all the other camels on the desert plain, some loaded up and some not. He thinks to himself, "I'd like to not have all this baggage on me for a while – I want to be able to explore the world a little and see what's out there." He throws off his heavy load and heads out into the world for some fun and adventure. However, it's not that easy. Suddenly he is confronted by a dragon who stops him in his tracks. This dragon represents *thou shalt* and the demands of society – the common thinking that it is not okay to give up everything and walk away, and that you must fulfil the role that has been planned-out for you. The dragon demands that the camel put the bags on his back and get on with his work. The camel, knowing that transformation is required to overcome the dragon, turns into a lion and slays the beast. He is then free to explore the world and discover what other options might be out there for him. The moral of the story is that it can be very difficult and challenging to step out of the mould that society has put us into. It takes courage, power, determination, and the strength of a lion to step outside these expectations and to go our own way in life.

Myth #5: Don't Make Your Love Your Work

Another myth was presented to me just yesterday – the perception that if you make what you love your work you'll end up not loving it any more. So, for example, if you love painting and then you start doing it full time for a job, maybe you'll end up hating it. Now, I can see how that possibility might occur. It is possible that if you spend too much time on something you might get fed up with it. However, this just highlights the need for constant vigilance around balance, the importance of taking breaks, and ensuring that what you are doing is giving you energy and not taking it away. However, finding your passion, and focusing your energy on that, will ensure you become an expert in that area, and that you are energized by doing something that you love.

Perhaps we have too many negative connotations associated with the word *work*. It can be used as a noun – activity involving mental or physical effort done to achieve a result. Or as a verb – be engaged in physical or mental activity to achieve a result; do work. Doesn't work sound much better through the lens of those definitions? When you think about it in terms of achieving results, it seems so much more appealing. To be honest, both definitions are much better than I imagined they would be – I think that I too may have caught the bug that makes you think anything to do with work is BAD!

However, I consider finding your purpose to be the possibility of finding a physical or mental activity that gives you energy, enables you to achieve a result, and allows you to earn a living. And I know, having talked to numerous people who have done it, that you *can* find work that is your passion. You will see real-life examples in the pages that follow.

4

TRUE STORIES

I have been lucky enough to meet many people over the past couple of years who are living their life on purpose. They have done whatever it takes to find their dream life, and they are living it. I interviewed some of these wonderful people either face to face or by Skype. (You can find the full interviews on my YouTube channel – the link is at the bottom of the homepage at www.sarahoflaherty.com).

I have amended the interviews to remove superfluous words, and any content that you might not find useful. I have also included a summary of what I feel are the key learnings from each person. Also, in each interview I am tagged as 'S' and the person I am interviewing is tagged with the first letter of their name. Anything in italics is my own addition while writing. I hope you enjoy these life tales from people who are not much different from you.

AMANDA – SHINE FROM WITHIN

S: I'm here with Amanda Rootsey, who established Shine From Within, giving girls the tools and confidence to shine bright. So, Amanda, we'll just start out with a little bit of history about you – your background, and what brought you to where you are now.

A: I did a business degree at university. While I was studying, I was also modelling and working in deportment schools, so going around to schools teaching young girls posture and how to present themselves like ladies. I was also working in hospitality. I was always pushing myself hard, I was a good employee. If someone called in sick, I'd be there. And then when I finished university I went overseas for a while. I modelled in Europe, and, while I was over there, I found a lump on my neck. It was just, sort of, there, and I didn't think anything of it. But eventually, when I had returned home, I saw a doctor and it ended up being Hodgkin's Lymphoma, Stage 4, which is a cancer of the lymphatic system. And that diagnosis shook me. I was twenty-four, and just didn't expect to be told something like that, as you can probably imagine. It was that diagnosis that totally changed my life, and amazingly, in the best way possible.

I learned so much about natural health, and how important it is to make the most of every day, and the importance of looking after this body of ours. I had two years of treating it naturally, as well as doing chemotherapy and radiation, and being rushed to emergency a few times, and, you know, it was a full-on time. And once that all finished, my partner and I just needed to recover for a while. My partner was amazing throughout that whole time. We ended up getting a shipping container cabin and putting it on a block of land that we found on Gumtree for $50 a week, and we stayed there for a year, growing our own veggies and being totally off the grid. It was such a beautiful time to recover, but also just to reconnect and remember what was important in life for us.

S: It's fantastic that you took the time to do that. I know when I first moved to Australia, I was interested in all those small homes and living off the grid, it's very rare to meet someone that's actually done it.

A: It was very romantic, I must admit. But at the end of the year we were ready for hot showers and being a little bit less removed. We had to put the four-wheel drive in low gear just to get there, so we were busting tyres all the time. By the end we were ready to finish up. But at the start, it was super-romantic. Everything you imagine it would be.

S: Yes, it sounds interesting. It would have been a good way to reconnect: to yourself, to nature, to everything, I imagine.

A: Yes. And then I started getting my energy back and I felt like I wanted to be out in the world a bit more again. Also, I just kept having these ideas come to me about running programs for teens, in a similar way to what I was teaching before in schools, but bringing in a lot more about natural health and self-love and positive body image and nutrition and wellness, and that kind of thing. So that's where the Shine From Within idea came from.

S: How did you first get it going? What were the first steps you took?

A: Good question. I think it was probably a couple of years of idea gathering and doing extra trainings, which I'm always doing, to be honest. I think it's important to always keep on learning. And then I just started writing the curriculum, I think. Putting the curriculum together, getting all those ideas down more clearly. Just taking it one step at a time. Eventually I started connecting with community groups that work with youth already, so I could start to find a bit of a market I suppose. I did a bit of volunteering around the place as well, and eventually ran my first workshop. And then I did things like building a website and developing a social media presence and connecting with my audience.

S: Right, and you started growing from there. Great. I know all the information is on your website, but do you just want to run through, briefly, what the different programs are that you offer? I was quite surprised, when we spoke earlier, by how many different options you have on offer. It's a lot.

A: Yes. Okay, so with Shine From Within we focus on tweens and teens, and the signature program is a five-day school holiday program for thirteen- to seventeen-year-olds. We have a tweens program for ten- to twelve-year-olds, which is two half-days. But we also run school-based programs, in primary schools and high schools. Sometimes they are after school, run as after-school programs. Sometimes the schools will get us in as part of their curriculum and as part of doing something for the whole grade. We run sisterhood evenings regularly, so once a fortnight we bring a bunch of girls together, just to chat. It's an opportunity for them to talk about whatever they need to talk about and connect with other girls, so that they're getting out of their regular friendship circles and families and school. It means they've just got that extra bit of support somewhere else. There's been a lot of research around peer support, and how that can really help to build the girls' self-esteem. We do one-on-one mentoring and coaching as well, and we do mother-daughter events too. We've got some mother-daughter evenings coming up and a retreat happening in Noosa as well.

S: Fantastic.

A: I think that's it.

S: Great. And do you want to talk about your new project that's coming soon?

A: Sure, so this is the Shine From Within youth mentor training program. It's for women that want to work with teens, because I get so many people contacting me saying they want to do something similar in their area. Or, the girls in their area really need some support, and they just don't really know where to start. There can never be too much of this kind of support for girls. The more programs out there, the better. So I'll be teaching the women that do this online program to get in touch with their strengths and values so they can create something that's unique to them and that's really true for them. Because that's how they'll connect with the students. The program will go through duty of care and the psychology of teens, so that they're well equipped to know when a student needs to be referred on to a counsellor or a psychologist. The program will help them to feel comfortable holding the space for teens. And it also includes business events and the marketing side of it as well, so they can make sure they're reaching the girls that they want to reach.

S: So when can people expect to hear about it? Start signing up for it?

A: Well, they can go to the website right now actually, and register their interest. There's a little button at the top that says Shine Teacher Training, so they can jump on the list now so that they're the first ones to hear about it. It's at http://www.shine-fromwithin.com.au.

S: Great, so I have just a few more questions. My book is about people who follow their dreams and who live on purpose and who are doing positive things in the world, and you'll be an amazing inspiration to a lot of people.

A: Thank you.

S: One of the questions that people often ask is, who's

supporting you? Who do you lean on? Do you have a mentor? What kind of support network do you have? And what would you recommend others have?

A: I have a little mastermind group I'm part of, and we have a conference call once a week, just for an hour on a Monday afternoon, to check in with each other and let each other know what we've got planned for the week and keep each other accountable. Because when you're doing your own thing, you can feel quite isolated, which I imagine is why people want to hear about the support networks that are in place.

S: Yes, absolutely. People want to do their own thing, and set up their own business, but it can be quite lonely, and they want to know, how do I get the social interaction happening? So how did you find the people that are in your mastermind group?

A: We were part of an online course together a few years ago, so we connected that way. Our businesses are in similar stages, they're not in the same industries at all, and I don't think that's really necessary. But it's just nice to talk to someone that also has their own business and are dealing with similar things. And we'll get together once a month usually to do a big picture planning session, and once a quarter to plan the next three months.

S: Great. That's very well set up. I've heard a lot of different people talk about mastermind groups and I've always thought it's a good idea but it's not as easy as it sounds to get working well. One of the challenges I've found is keeping it going consistently and making sure you find the right people. So yes, it's good that you've got a good group.

A: Yes, it's quite a small group. There's only three of us, so that works really well, without having too many elements to disrupt it and cause people to be unavailable, and that sort of thing. But, people do sometimes outgrow their mastermind groups, in time, and then they can start looking and searching for something else.

S: And any other support structures?

A: I love coaching. I'm not with a coach at the moment but I really believe in the power of coaching and having that extra support as well. So I'll quite often have a coach. My partner Dave, he's really supportive too. He runs his own business, as well, from home. It's nice to have that understanding from him. And it does mean that we're both at home together a lot of the time, so we don't feel so isolated.

S: You're not totally by yourself.

A: Yes. And I do have some great girls that help out with the courses and provide administrative assistance. So I feel very supported having them there as well.

S: Are those girls that did your courses? Or are they from outside of that, that you found in other ways?

A: A couple of times I have had past students come on board to do work experience, or paid work, actually. But I'm finding, at the moment, that I'm really enjoying having someone a little bit older, that's got a bit more work experience behind them. If only for the fact that a lot of the work that I'm asking them to do has to be done on their own at home, and it's hard to motivate a teen, who hasn't had a job before, to do their own work from home. They want to be in a fun work environment. I know I certainly did, back in the day.

S: Yes, absolutely, that totally makes sense. Okay, so back to you. Overall, what do you think the benefits have been to you, for you? You were on a certain journey where you were doing modelling and many other things, and you could have got a job for a company or followed a traditional work pathway. But you chose quite a challenging direction, because you're doing something that's new, and that you've got to set up and do by yourself. Obviously, you wouldn't have chosen this direction unless you felt strongly about it, so what pushed you to go that way? And then, what do you think the ultimate benefits have been?

A: I guess what pushed me to go that way, as you said, was believing that it was important and that I really felt called to do it. You know, these ideas were just coming at me all the time. And, there were a couple of times over the last five years or so that I've had to go and get a part-time job as well, and in my mind I was thinking, you know, it would be so much easier to just get a job and get paid to do it, and then come home. That'd be great. But whenever I've tried that, I find that I end up working on my own projects as well, and so I'm trying to fit in a whole lot more. Those times reminded me that I really do love working this way. And even if I'm working on other projects, I still always come back to doing this as well – this is what I really want to be doing.

S: Yes, I totally understand that. I've had that same experience. You choose a direction with your own business, and then you get some paid work, and you think life would be easy if I just did this. But then you know that you couldn't just do that, it wouldn't be satisfying.

A: And I don't think there's anything wrong with that either. I think a lot of entrepreneurs will speak badly about having to get a job for somebody else. Or supporting somebody else in their dream, as though the only ultimate option is to be working on your own and making six figures, and all these things. But I think it can be valuable to work for someone else as well, even just to get out of your own headspace and to be working in someone else's business, and get a different perspective. Yes, I don't think it's a bad thing.

S: No, I think you're right, because you also shift your mindset a little bit, and you're interacting with different people. You're in a different role. For example, I work part-time in a customer service role, and it puts me in quite a different headspace than I am normally. I think it helps to shift my perspective, and you hear different things from the people you're interacting with. You're dealing with all these people that you might not normally be dealing with.

A: Yes, exactly.

S: Yes, you're right. And I think it shows a lot of flexibility too, that you're not so rigid. Maybe that's one of the reasons why some entrepreneurs succeed. The ones that succeed, I suspect they don't just go, I'm going to do this and I'm going to succeed in six months, and that's it. Because that's not how it works – it takes time and flexibility.

A: Yes, so true.

S: It takes a little bit of time, and effort, and flexibility. A lot of people often won't go it alone because they just feel like it's going to be impossible for them to make money, to be able to pay their mortgage, and all that sort of thing. So I guess, if you could just talk about that a little bit, maybe allay some concerns around this subject.

A: Yes. I think that's just fear creeping in, isn't it? And it's hard for me to comment too much about that, because I don't have children that I'm supporting, I'm very aware of that. But, for me, I've just found that I always am supported. I certainly stress about money at times, that's for sure, but then something comes, and it always comes, and so I've never been in a situation where I can't pay my rent or anything like that. I've always felt supported. I certainly get tested sometimes, to find that trust and that faith again, because it can be so up and down. But, yes, I think that's just an excuse though, because you can make money doing anything.

S: Yes. I think if you're doing the right thing, you know in your heart it's the right thing, and it helps to give you the faith, the faith and the confidence to keep going. I'm a big Tim Ferriss fan, so I listen to a lot of his podcasts, and he quite often says, what's the worst that can happen? Can you imagine yourself living on the street? Can you imagine yourself having no food? And I think, well no, not really. I can't imagine it getting that bad.

A: Yes, yes, exactly.

S: I wonder what we're so worried about sometimes.

A: Yes. And, I think that maybe a big part of it is also having that flexibility. I imagine if you had in your mind that it must be this way, and I must make this much money doing this exact thing, and if you weren't willing to budge a little bit, or feel a little bit flexible, then it would be very scary.

S: Scary. Yes, you're right. I think that sort of tension or rigidity can cause a lot of stress and fear. That just reminded me of something we talked about a little bit last week. I remember you saying that you do public speaking sometimes. You call yourself a youth presenter, and you go to youth events and present with other people. And what impressed me about our conversation was that you said that a lot of the youth presenters are very extrovert, and very outgoing, and quite loud. And yet, you'd had people comment to you that they were impressed with the difference in the way you presented compared to those people. So do you want to talk about that a little bit? Because I think that comes back to the fact that you're quite clear on who you are and what you stand for. You're not being influenced by how other people are being, which I think is important. It's so easy to feel like we have to be like everyone else sometimes, and in the process we lose ourselves.

A: Yes, and I certainly was very much like that, to be honest. You know, I'd be comparing myself to someone else who is super-vibrant and gets super-energized from being in front of the crowd, and is funny, and all of those things. And I'm just not those things. I can try to be that, but I run out of energy about five minutes into it. I'm slowly realizing that it's okay to be quite gentle, and it was nice to get the feedback that people enjoyed the calm energy that I brought to balance things out. So I just think it's important to remember that whatever you have is okay.

S: That word, gentleness, you use a lot. That word keeps resonating with me, I think because it's quite unique. It presents a soft energy, and it feels like it's what a lot of people probably need at the moment, it's very authentic. It's so nice to meet someone who's quite clear on what they stand for, and who hasn't been swept up in all the other things, you know, the images and what people feel like they should be like.

A: Yes, thank you. I certainly get swept up in it every now and then, that's for sure.

S: Another thing that you talked about was the eco-modelling you do. You mentioned that you are quite picky about who you do modelling for now.

A: Yes. After going through that whole health experience and coming out the other side, I just didn't think I would get back into it at all, to be honest. It felt a bit frivolous to get back into modelling. But, I also didn't like the way that many in the fashion industry do business. I remember putting on a Salvatore Ferragamo fur coat, and realizing it was an animal I was putting on. While I kept on going in that moment, I've always remembered that situation. I wasn't going to get back into modelling at all, but a couple of eco-friendly labels got in touch and wanted to work with me, and it grew from there. So I'm Australia's first eco-model and I work exclusively for eco-friendly, ethical, and vegan businesses. And I write about eco-fashion for *Nature & Health* magazine, and I'm on the board of a couple of sustainable fashion groups. I think it's so important for people to be aware of that side of things too, because we all have to get dressed in the morning. I think a lot of people go, I'm not really into fashion, so I don't really care about eco-fashion. However, no-one goes out of the house naked, so we should be taking an interest in it, the way that we take an interest in organic food on our plate.

S: It is important, and I think a lot of people feel there's so much wrong with the world that they can't possibly make a difference. But I always think, how you spend your money is how you make a difference. It's what clothes you choose to buy, what food you choose to buy, etc. If everyone lived with that mindset then we would all be making a huge difference.

A: Yes, definitely.

S: I think it's fantastic that you are involved in all those projects, and in an industry that's clearly growing.

A: Yes, and it comes from, as you said, people voting with

their dollar. And, asking the right questions when they go into a shop – who made this, what is it made from?

S: Yes. So when you talk about eco-friendly fashion, what does that mean exactly?

A: That's a good question, because it can mean anything, especially when you start to use the word 'ethical' as well. And it can mean something different for everyone. For me, I look out for how it was produced, the sort of fabrics that have been used; and I look for natural fabrics as much as possible, and organic. Because, cotton, the way that it's sprayed conventionally uses a ridiculous amount of pesticide, and it affects waterways, animals, and even the people who are spraying the cotton get very sick as well. So looking for natural fabrics, looking for Fair Trade, or something along those lines. There's still over two million child slave labourers in the world, and even in Australia, there can be instances of that as well. Ethical Clothing Australia is a great symbol to look out for. If they've approved a brand, then it means the people that are making the products are at least paid minimum wage. There have been a few instances where immigrants, in particular, are being exploited, essentially because they don't speak English. Even here, right under our noses.

S: Yes, that's quite scary, isn't it? You always imagine that's something that's happening in China, as everyone says, and you don't think that it could be happening in Australia.

A: Yes. And, right down to the way clothing is dyed. There are some great labels, like Sinerji Fashion, that use all natural dyes, as well as vegetable-based dyes. You can find images online of waterways in China that are just bright purple, and that's how some fashion labels will figure out what the trend is going to be for next season. They check out what's happening in the waterways, and realize that that's the next season's popular colour.

S: Wow, scary.

A: And for me, also, I make sure everything's vegan. I don't like to wear anything that's exploited an animal. So, for me, that means no wool, no silk, no leather. Yes, that's it I think. No fur.

S: Yes. I think I heard that Natalie Portman does that too, which I always thought was impressive. So what do you wear? And footwear-wise, what kind of shoes do you normally wear?

A: There's a great company in Melbourne, called Vegan Style, and they have some great options. I've got lots of their shoes. A brand called Melissa do some great shoes out of recycled plastic, which are fun. And this is where the debate tends to come up too – a lot of people feel that they would much prefer to wear leather than to wear fake leather or PVC, because that's made from petrol and the way that it's produced isn't great either. So it just comes down to what is important to you, and being careful about only buying what you need, and looking for it second-hand, and those sorts of things as well.

S: Yes. I think, again, a bit of flexibility is needed. You know, it's impossible to be perfect, and I think recycled plastic is better than some other things. Thank you for sharing your knowledge.

A: Thank you.

S: Looking back over the last little while, is there anything you would have done differently? Or is there anything you would share now with the you that was just starting out?

A: I wouldn't change anything. I think I would just say keep going, really. Just take it one step at a time, because I've found over the last few years I've got overwhelmed quite a bit. But it always works out, and yes, if you just take it one step at a time you'll be okay.

S: Good advice. And it's quite interesting you say that, because running a business on your own, it's a lot of work, isn't it? How do you manage not to get overwhelmed?

A: I find having those check-ins with my little mastermind group help. And simple things like sitting down and writing out my three MITs, or most important tasks, for that session. Because otherwise my to-do list can be two pages long, and then I just get scattered and it's too much and I freak out. So I find I must be pretty disciplined and make sure that I write down just three things that I need to get done. And then, if I get through them, I'll do another three. It helps to get really clear on what your priorities are, and I love utilizing different apps too. I love asana.com, that's a great app for keeping track of what you need to do, and breaking things up into projects. You can access everything on all your devices, so if you think of something when you're out and about, you can just add it in there. It's a free app. And, of course, there are things like Evernote that are great for keeping stuff together, and even Google Docs I'm finding helpful just to keep everything in one place.

S: Yes. I must admit I use Evernote and Google Docs, they're both fantastic. Now, I think you were going to share a few confidence tips?

A: Yes, so I taught a course this morning, for some teen girls, and we talked a bit about self-worth and confidence. Some of the things we discussed were: be true to yourself, keep it simple, and being okay with who you are. These three key things are what it comes down to for me.

S: That's great. Isn't it funny? It's so straightforward and simple, but such good advice.

A: I find I learn so much from the young girls. They have these moments of such clear wisdom, and it is usually so simple. I think, as adults, we complicate things and feel like we must read five books on confidence instead of just going, you know what, I'm okay. I'm okay the way I am. I'm doing okay. If I be a bit gentle with myself, I'll be alright.

S: Exactly. That is so true.

A: And tips for starting out? I would say find a group that you can connect with, because it does make all the difference, and, for me, that's meant enrolling in different online courses. Because with that usually comes a Facebook group where you can connect, and you're all at a similar stage, so you start to build your business together and can ask each other questions. I find that really useful. But there are also lots of great Facebook groups for women entrepreneurs, and for health coaches, and for all sorts of different people. I would recommend finding a tribe for yourself that you can connect to. Because, especially if you can find a couple of people in your own area, like we have, you can have a coffee with them, and just feel like a normal person again. Like, you're not going crazy working on your own.

S: Someone else to connect with. Thank you so much for your time. On Amanda's site, I know that you can find out how to work with Amanda, information about Shine From Within, eco-fashion, simple living, and information about vegan food as well.

A: Thank you. Thank you very much for having me.

Key Learnings:

- Ongoing learning and development is important, and don't be afraid to volunteer to get the experience you need.
- Establish a strong support network. Consider setting up a mastermind group. This kind of group can help to keep you on track to complete your tasks and meet your goals. Consider finding a coach as an additional support.
- Find your tribe – a community of like-minded people to connect with. These people may be able to help you and your business grow.
- Notice what's calling you, what you feel driven to do. What do you keep doing whether you're working in another job or not?
- There's nothing wrong with getting a part-time job, if and when you need to. The exposure to other people,

and other businesses, may assist in your own and your business's development.

- Be flexible.
- Do what fits with your values. Make sure you know what your values are. *You can find a list of values on my website at sarahoflaherty.com.*
- Be true to yourself, be okay with who you are, and keep things simple.
- Keep going and take things one step at a time.
- Take some time to write down your most important tasks. Try for three key things each day. Utilize productivity apps to help you manage your workload.

You can find Amanda at:
www.shinefromwithin.com.au
www.amandarootsey.com.au

RAM – FOUNDER OF INNR.ME

S: Let me introduce you to Ram. Ram is the founder of a global wellness coaching platform, called Innr.me. Welcome Ram and thank you for joining me.

R: Hi Sarah, thank you for having me on board, I really appreciate it.

S: No problem. So before we start talking about your background and how you've got to where you are now, do you want to just talk a little bit about Innr.me and what it is so people can understand why I'm talking to you?

R: Yes, that would be great. So Innr.me (the website is www.innr.me) is a global wellness coaching platform. What we have done is create a platform where people can connect with coaches and experiences from all over the world, to learn from them, and to live well and thrive. There was an opportunity we saw that there were coaches from everywhere who were offering amazing wisdom and coaching in different aspects of people's lives, but there wasn't one place where people could find everyone together, and Innr.me does that. Innr.me basically brings together coaches from multiple wellness modalities onto a single platform. And we've built a way where we bring the best of wellness coaching and the best of technology together to make it super-easy for people to take control of their wellness journey and connect with coaches, and learn from them, and live an awesome life. That's what Innr.me is.

S: It's really exciting and such a great idea. So tell me a little bit about yourself. It's always good to know where people have come from to get to where they are now.

R: Well, I'm going to tell you a long story then.

S: Okay. Great.

R: Just immediately prior to Innr.me, I was an advertising professional. So I was in the advertising, marketing, digital, you know, media world for over a decade. Prior to that I began my career as a technology consultant. So I am an engineer by training and an advertiser by profession. These two careers have really shaped me, by bringing out the best of what my college schooling had to offer and the best of what my marketing experience had to offer. Now personally, I am from India. I grew up in a very small town in southern India and I am the very first guy to finish high school in my family. The only reason I say that is because nobody gave me career guidance, which is maybe why my career has taken a very interesting turn recently and has been an amazing journey over the years. And, another reason why I want to share that, is because I started out as a typical South Indian boy, where everybody dreams of becoming an engineer. So that's where I started out, but I had other dreams. I wanted to explore the world, and I really wanted to be in New York for some reason. Frankly speaking, New York was so far away for me, the moon seemed closer. And so one thing led to the other in my life, but there was always this desire to explore the world. Once I finished engineering, I ended up getting into a university in Texas – Texas Tech. I got my masters in management information systems and got a job with Ernst & Young, which is a large consulting company. And that's really where this whole journey started, because about a week into my very first high-paying, amazing job I realized that something was off. It took me another ten years to understand that there was something really off. To start with, I was a highly successful business analyst for Ernst & Young, where I worked with large companies, such as Walt Disney, Texas Instruments, implementing their customer relationship management strategy. However, something was missing, and I want to share one little short story about my time with Walt Disney.

S: Okay. Ram, sorry, just quickly, before you tell that story, can I jump back to when you went to Texas. You said you were in South India and no one else in your family had been to high school let alone university. And yet, you ended up going to a university in the States. That sounds like a massive move. How did you manage to make that happen?

R: It was a combination of things. When I look back I believe it was the law of attraction. I had no money, and I had no way of going there, and yet there was something in me that completely believed that I needed to be there. When I was in my eighth grade at high school in this little small town, we used to get one magazine from America, which was *National Geographic*. There was an issue about New York and it included a map. For some reason, I tore that map out because I loved it so much, and I put it in my bedroom and I memorized the entire grid of Manhattan.

S: Oh, that's amazing.

R: Yes. I completely memorized it. I would dream about it. There was no sense of doubt or scarcity or fear that I had no money to get there. And somehow or other, everything came together, and the next thing I knew I got a scholarship to Texas Tech and I was there, you know? And so, honestly speaking, I do not know how I made it happen, except for that complete, no doubt, sense of energy inside me which just pushed me and it got me the right people to help me. Without the right people I would not have got there. And even once I got there, people helped me, and I was able to be open and tell people what I needed. I spoke to my professors, I spoke to people around me, I got the right kind of jobs and opportunities, and it drove me towards what I wanted. And, the next thing you know, fast forward, the very first time I ever actually landed in New York I did not need a map. I literally did not need a map. I walked around. I'm like this is it, this is it, this is it, this is it, and it was a very surreal experience. So, to answer your question, it was the strong desire and the belief that I will one day be there which I believe took me there.

S: Yes. That's amazing. I guess it's how people are sometimes when they're young, they don't realize that they might have challenges or barriers. You can't imagine anything else. There's no fear. There's no barriers. And it happened.

R: Yes, I sometimes wonder where that attitude is now. Because I wish I had that same sense of wonder and the mindset that I'm going to go get it no matter what it takes, you know? And, I try, even now, to anchor back to that younger Ram, and say hey, you've done this before, it's alright. Because as we get older, as you might relate to, we lose that sense of unabashed positivity, that sense of let's just gun for it.

S: Yes, as we get older we get a bit more fearful, and worry about things like am I going to make money, and I need to make sure I can pay my mortgage. And all these things can get in the way. I'll take you back now ...you were going to say something about Walt Disney I think?

R: Yes. So I spent about six months in Walt Disney World in Orlando, working in their business, and what I learned was that the entire multi-billion-dollar business rests on that little mouse. Everything rests on that little mouse. That was when I understood the power of branding. Everything that you ever see in the entire Walt Disney experience is focused on the mouse. So, that got me thinking, wait a minute, I'm working on creating this amazing platform where people can buy tickets and other things, but then, what's it all for? To experience the brand, to experience what it promises. And, I said to myself, "Okay, I want to learn how to build brands, create brands, and help brands build better connections with their customers." So, that was when I started investigating branding and brand development, and I discovered that it came naturally to me. People would always say Ram, you're great with ideas, and I thought, I want to pursue this as a career. You know, firstly, I wanted to be around people who are in this work, doing branding and advertising and marketing, so I can learn from them, and secondly, I wanted to get to a place where I could try doing it for myself. So I decided to switch my career, and even though I had no idea what marketing was, and I had never done advertising before. I said, okay, if I'm going to be in America, I'm going to be in advertising. So, I just quit my job and pursued that.

S: Wow. That was brave.

R: It was scary. It was crazy. And I think all my well-meaning friends said, how are you going to take care of your visa? You know, what are you going to do? You don't have any experience. You have a great paying job with Ernst & Young. And for some reason, something told me that it was more painful for me to continue in a disconnected environment than try and take that leap into something new. And again, I jumped in, I got in my car, I packed everything, drove from Dallas to New York and ended up at a buddy of mine's place. And I said, okay, if I'm going to be in advertising, I'm going to be in advertising in New York, otherwise there's no point. I had no idea. I didn't have any connections. I ended up in New York, and New York literally changed me, and opened my mind, and humbled me. Most importantly, it humbled me, you know, it grounded me, and it also gave me an opportunity to work in a great agency in New York City.

S: So, yes, what I was going to ask is what agency were you working for when you started out in New York?

R: I started out with a small little boutique agency, called Jesse James Creative, and it's still around. But I was gunning for a large global agency and my second job was with Publicis, which is one of the large advertising agencies out there. I was the second hire for Publicis Dialog's interactive division, that they were just establishing. And I must let you know this, again I point back to the belief that I had, because many of the people trying for these jobs were more qualified than me, they were smarter than me, they had a better degree. And yet, they were not able to get into an advertising agency. Here I was with no degree in marketing or advertising, and no experience, and I was able to get into that agency and convince them to please give me a chance. And they believed in me and took a chance.

S: I've always thought of advertising as business consulting, so you came from the right background in a way. And I'm sure they loved that you were an idea generator. From their perspective, I'm sure they knew they'd got a great hire, even though you didn't have the typical background.

R: Yes, thank you. I was given the opportunity to learn from all these amazing, amazing, business leaders. And they gave me the opportunity to work with some of the top brands in the world. My engineering background helped a lot. There were creatives coming up with amazing ideas that were really high up – ten thousand feet level – and with my engineering background I was able to bring their ideas down to the nuts and bolts, this is how we execute level, and that is really where I brought in my unique experience. And that experience was really appreciated by the clients, as well as by my peers.

S: So you've got the engineering experience, you've ended up in New York, which is where you wanted to be, you're now in advertising. So how long did you work in advertising for?

R: I was in New York for about five years. Then I left New York because I wanted to get back to India. I spent from 2008 until 2011 in India. Because I had never worked in India before, it was like coming into a completely new environment, a new culture. Personally, I was familiar with where I'm from, but then, professionally, I had no idea. For the first six months I wanted to get on a plane and run back to America. But I realized that there was an opportunity here. I mean, I think Asia was changing, India was changing, and I had to understand and learn if I was to become a global professional. It was something that I thought would be a great opportunity. And being in India grounded me, and gave me the opportunity to work on the same kind of brands I had worked on in the past but in a different environment. I had the opportunity to work with Coke, I launched Sony's Expedia range of smartphones, I did a lot of branding work for Avia. So it brought me to a place of, how do I take a global idea, a global brand, and make it relevant to a market which is not as highly developed as a western market?

S: It must have been interesting going from a market like America to India's. I mean, really, totally different markets, and totally different working environments.

R: That's right, and that's what led me to the next opportunity, which is where the genesis of Innr.me comes in. I had the opportunity to go to Bangkok, and I think that's where we connected. [S: I first met Ram in Bangkok, where, at the time, we were both working for big-name advertising agencies.] I spent a couple of years building a great team and implementing digital media for the Interpublic Group Media brands unit. And that's when my journey with Innr.me started. What I realized was that while my career was taking off and it was in a very, very good place, I felt empty inside. It was a very funny feeling for my male Indian ego, you know. My ego was fighting with me and saying, hey, you know, you are this, you are that, this is great. And I was very grateful for the experience, and I was very happy about the professional accomplishment. I still am, and I'm very proud of it, but yet, there was this nagging feeling that something was amiss, you know? And while I was working with all these top brands and celebrating and learning, I realized that I wasn't sure that I was living on purpose. I wondered, am I in my alignment with who I really am? And it took me a long time to admit that there was something amiss, again. You know? And that is where this eleven-year journey took me. All the way from me getting into Interpublic Group's IPG Media brands unit in New York, being on the top of the world, ending up in Bangkok doing some amazing work – launching Windows 8 and having the best digital media campaign for Microsoft globally – and yet, realizing there was something amiss. What was it? Am I being authentic? And, can I just be honest with myself? You know? And one thing led to the other, and I admitted that things weren't right, and I started asking for help. Reaching out and asking for help. And that by itself was a big step for me.

S: As you can imagine, that story resonates with me. Being very successful in a career, and thinking, I can't throw this away. I'm really lucky to have had this successful career and I've worked really hard for it, and it's what society thinks is successful, and yet you just know it's not right. And, as you said, you kind of feel empty inside. It's like you're not being authentic to yourself. So, yes, I totally relate to that. Then you mentioned you started asking for help, can you explain what you mean by that?

R: Yes, that's a very good question, Sarah, because I didn't know what I was asking. I didn't know what the formula of the question should be. I was so disconnected from my own self and what I was looking for, that I was just, kind of, focusing on my symptoms – I'm stressed, I'm feeling this, I'm not happy. It was all very top-level questions that I was asking initially. So it became a process. The challenge that I had was to articulate what exactly was the challenge that I was experiencing. And even for that I needed to reach out for help. And, you know, especially where I'm from, men don't just say "I have a problem." So that was one very challenging aspect that I had to fix. I thought it was an expression of my weakness, there was something wrong with me, and that other people would judge me. It was all a very secretive process at first, right. So I'll go and present this great strategy to some great group of clients who are all awesome, but then, when

I'm at home, I'll feel miserable and I'll be Googling my way through, what happens if you're alone. And it was a very secretive process, because I was ashamed of myself. I still remember once when I was at a regional conference out in Singapore, and we were all partying and so on, you know how pleasant folks are in our industry, we are happy, we are having a great time, and yet I came back to my hotel room and I cried my eyes out. I was thinking, I am not well, this is not right. And I'll admit it, it was very painful. And I said, wait a minute, that group of people out there love me, they're all great colleagues, why am I feeling bad? Why am I feeling disconnected? And then that night I decided that I needed to do something about it. So what I did was to reach out to as many people as I could and ask...literally writing out the kind of questions that I wanted to ask...and I started asking those questions. And, thankfully, I was able to connect with some amazing coaches. And I do believe that a complete specific shift happened inside me, because I think the universe was just waiting for me to just say, admit it, you need help. And the moment I said, yes, I started connecting with all these coaches, I learnt from them, got coached by them. That was where I sensed the opportunity of Innr.me.

S: Thank you for sharing your story – I appreciate your courage in sharing what was obviously a very tough time for you. I'm sure there are lots of other people that have felt the same, and will appreciate your story. And, as you mentioned, especially men struggle so much sometimes because guys don't talk about stuff the same way women do, so it's often more challenging for them. And when you are in a successful position, it's all about professionalism, and it's hard to share, especially when you're not feeling great about things.

R: Even more funny is how the universe plays to you, because I was in the land of smiles, Thailand, and I was miserable. And that doesn't make any sense, right? So I'm in Thailand, everybody's happy, but then I realize it's all within me. The things that came up for me were, let me change my job, or let me move out from here and go back to America, or other external things. But none of that was true, the only way out was in. And it was a very humbling process for me to get there. But it was also the greatest opportunity presented to me, and I had to go through that shift to understand it. And then I said, okay, there is an opportunity for me to, first of all, learn and grow and become who I really am. Which also gave me the opportunity to realize that I'm not alone in this. There are hundreds and thousands and millions of people out there who are on the same journey. How do I create an opportunity where I can provide the same kind of access which I have had, and make it easier, you know? Rather than being all alone, being in the dark, not being able to share. Let me create this opportunity where it can be easier for other people to go through this journey. And that's really what Innr.me is.

S: It's very interesting to hear you tell your story, because it's almost like what you learnt very clearly showed you what you needed to offer.

R: Yes.

S: Yes, and I think that's what happens to many of us. We go through some challenge that is the universe telling you you're not on the right path. So we go through a challenge that then shows us where we should be. It's very interesting what you say about happiness comes from within. You can be in the land of smiles, but it doesn't matter where you are, it's what's going on inside you that matters.

R: Amen. Absolutely.

S: Well, great story, thank you so much for sharing it. So, I've got a few other questions. What have been the biggest challenges that you've had to face setting up a global platform? I mean, it's a big deal, and a huge undertaking.

R: Sarah, to be very open with you, I think most of the challenges have been inside of me, and I think now, your readers might be able to relate to that. However, when it comes to me, working on a large brand, I have no doubts in my mind. So I can, given my training and my experience, sit in front of any brand of any stature and help them create a strategy for them to achieve their business goals. But the challenge for me was to apply that for myself, for a brand I was developing. The greatest barrier has been for me to, first and foremost, admit to the fact that there has been a sense of 'not-enoughness' when it comes to my own self. And number two was my hesitance to reach out for help. So these two were the biggest barriers. Now I'm facing the same level of challenges that any person would face jumping into creating something. Resources, access to capital, and that's standard, you know, entrepreneurial challenges. Which I believe, in my heart, are not a big deal. They're just business processes, which needed to be tackled at a business level. The barriers that I have faced, now I'm learning and growing, are completely within, and the biggest, single biggest barrier, has been not-enoughness.

S: Yes. That is such a good way of putting it. Not-enoughness. We were talking, before we started the interview, about different podcasts that we both listen to. And, I've been hearing that not-enoughness coming from famous people who are super-successful. So, you've got to accept that we've all got that in us, that we all sometimes don't think we're enough.

R: Yes, and I think I've learnt it. I think it's something I've overcome. However, for all I know, it might be there for the rest of my life, you know? But the fact is that I have recognized it. The entire journey has been about I'm not good enough to start a company, I'm not good enough to design this website, I'm not good enough to talk to a coach. Or I'm not good enough to go talk to an investor. Despite the fact that if I actually go out there and sit in front of a coach, I'm good enough to give them a strategy, I'm good enough to tell them what they need to do, you know? So it's learning and understanding how much of my life has been around outsider approval, outsider recognition, and outside reinforcement. And I wanted to understand this, and I've created this platform as the antidote for it, which is inside out.

S: Yes, absolutely. I know it's a challenge I've had to face as well, but once you're on the path and heading in the right direction, you just have to overcome any fears you have. You're on this road, and you must keep going and, then you're like, ah, okay, I can do this.

I listen to a lot of different people talking about their successes, and one interesting story I heard recently was from Sarah Blakely, who set up the Spanx brand. She mentioned that the best advice and support she'd ever received had been from her father, who told her that if you don't fail regularly you'll never succeed. I really like that piece of advice.

R: I just want to add one more thing to it, because I've faced the same situation. For me, I've had this experience where I get very inspired by the kind of experiences that people share. And I get it, and I resonate with it, but the challenge for me has been to fully get it at a heart level, at an energetic level, at a level where I believe in my bones and my cells and in my energy that it is true.

S: That's a nice way of putting it. It's all very well to listen to all these different people, but until you experience it yourself it doesn't really sink in.

R: That's right, yes. And I think that also alludes to the fact that we all have barriers. Sometimes these barriers won't let these new learnings in. And it's very hard to shift these barriers alone. That's where you need people who are qualified in helping you to open up and to guide you through this process of integrating this kind of wisdom that you need. And I hope that Innr.me will help you do that.

S: One thing I've noticed is that a lot of people who follow their dreams tend to be very busy. It's a lot of work setting up on your own, doing your own thing, and making sure you're being authentic. My question is, how do you fit everything into your life? Do you have a support network? Do you have go-to people?

Do you have mentors? You know, how's that working for you? How are you managing your work–life balance now?

R: That is honestly a work in progress, Sarah. I have not been very good at balancing everything in my life, especially after I started this whole journey. I think for the first six months I was in a very different place, because I was in a place of 'I must finish this.' It was all about deadlines. It was also a time where I was learning about myself, where my black spots were, etc. I made a decision about two or three months ago that I really need help and support. I had to admit that I cannot do it alone, and I do not want to be in this vacuum where I am trying to create this by myself. And the moment I admitted that is when I started connecting with people and friends, and sharing my challenges, and getting advice from people who have done fundraising before, who have worked with investors before. And the most challenging thing has been to talk to other people who are in the field of advertising and social media marketing. Because I want help in all other areas, but when it comes to marketing and advertising I am a Jedi. I was hot, right? So I'm still working on it. My balance is much better than what it was about six, seven months ago. I do have a support network, but that support network for me right now is mostly personal friends, because the greatest need I have is to feel that I am on the right path. I know that I am on the right path in my gut, but then, once in a while, doubt does show up. And I'm not going to lie to you, okay. We all go through that. And, for me, it is important for me to keep that anchor so that I can connect with my trusted loved ones, friends, family, and peers. But with respect to business, I reach out to people, I connect with them, and I ask them questions. But for me, the need has been mostly personal, and I do keep that in mind.

S: I think you're right and that's probably what I've had the most need for too. It's when the doubt comes up, it's important to have some good friends who will remind you that you're on the right track, and that you're doing well, and that sort of thing.

R: I would rather have those five great people in my life than have a thousand LinkedIn connections.

S: Now, can you share what you feel have been the biggest benefits for you of following your dreams? Has it benefitted your health, your mental wellbeing?

R: It has definitely benefitted me. The most important benefit that I have derived is to understand where all the opportunities are for me to grow. So given that I come from an extensive corporate background, I thought that I could manage this entrepreneurial journey very easily. It was not the case. The benefit has been to show me the blank spots where I can still learn. This journey has provided me with multiple instances where I was told by the universe that you are on the right path. And every time that happened, I knew it meant to follow your heart, go through this, face your fears, you know. Find your courage within you and go for it. The challenge for me has been to go to places that I have been avoiding, and this has given me that opportunity. And for that I am eternally grateful.

S: Have you had any experiences of synchronicity, because I know that when you start on the right path, even though you have challenges, and you may have these 'I want to eject' moments, quite often something will happen that tells you you're on the right path.

R: That's a whole different interview, because I can go on and on. I moved to Bangkok on 11/11/11. From the day I moved until now there have been so many synchronicities. And I'm only sharing this with you because you brought this up, okay? Some examples have been that every question I ever wanted the answer to, I would get an answer through a podcast, and at just the right time. There have been other things too, but I won't go into the stories here. But I have to tell you, if you had asked me about this four years ago...I'd be like, synchronicity, are you on drugs? But that Ram and this Ram are very different.

S: It's so amazing how, when you're on the right path, synchronicities do start happening and showing up in your life. Whereas when you're not on track, and this is what many people are facing if they're stuck in their jobs or something is holding them back, they face more challenges and the synchronicities aren't there; they're not showing up.

R: Yes. Three weeks ago, an ex-client of mine wanted me to check out this new, really high-profile job in a very large global technology company. Now, only because he recommended it, I went and connected with the hiring manager. Just to understand it. The job seemed great, the company seemed great, everything seemed good. But from when I got into the car, from the moment that I turned the radio on, every sign between my office and that office – the podcasts I listened to, the signs I saw, everything – was like, what the heck are you doing? Follow your heart. Why are you even doing this?

S: Yes. Well, it's tempting, isn't it? Like me, you used to earn a lot of money, and when you get offered a lot of money it seems like an easy, safe option, but it's not the rewarding option.

R: It really is not. But the meeting helped me realize that I am creating a collaboration platform. Why the hell would I go out there and do anything else? It makes no sense, you know? And while I was thankful for the opportunity, I was also thankful to the universe for giving me a very clear indication that I am on the right path. So I sent them an email saying, thank you for the opportunity. If you want me to consult and help you guys out, I'll be absolutely open to it. But my career path is very, very clear. But as I said, these kinds of synchronicities keep happening, and they keep giving me the opportunity to say, follow your path and follow your heart.

S: My next question is what traits do you think are important, or have been important for you, to follow your dreams? And I ask this because I think it will be helpful for other people to under-stand what characteristics are important.

R: I think self-honesty, authenticity, connecting to a deeper part of you. Coming from a place of power rather than from the head. The head is critical, however, the challenge for most people is to connect to your heart. This is where you will find true knowledge.

S: So self-honesty, authenticity, and connecting to your inner power.

R: Absolutely.

S: I just have a couple more questions. One of the barriers, I think, to people following their dreams is the financial aspect. A question that I ask everyone is, do you believe it's financially viable to follow your dreams? What's your perspective on that?

R: I can only speak for myself, and I can't advise other people because it's a very circumstantial thing. I believe it is viable. The challenge, I believe, is that people might not have thought about their life situation, and may be approaching their new direction with a pie-in-the-sky type approach. In my case, it's a different story. I'm single and I can fend for myself. But different people have different priorities. I have a friend who wants to quit her job at a large technology company because she hates it. But she has financial responsibilities, for her family. She has asked me about doing her own thing, and I can help her, but she needs to plan to do it over time, and transition it, so she can still look after her family. You need to ground yourself in your own reality, and to learn, and to seek out people who can help you plan your journey.

S: Yes. Good advice. So looking back, is there anything you would have done differently, or any particular advice you might give to your younger self?

R: Oh my God, so much. I think most importantly is to learn to love yourself. That's it.

S: Okay, that's nice. Do you have anything else you want to

share? Any advice that you'd give others that are considering starting out in business themselves?

R: You know, I do. I think we are all on this journey. I'm starting up Innr.me because I want to do this. I don't even want to call myself a founder, or an entrepreneur, or any other label. And I think we need to move away from identifying ourselves, from being labelled as something – for example, I am a start-up founder. I implore people to see if they are in a place where they're authentically having a wonderful time within themselves. And be honest with themselves. To move away from 'have to' and 'must do' and 'should', and say, I have one life. You know that maxim that it's so rare to have a human life, a billion, billion chance. It's one life, okay? And it's been so hard for me to even get that into my system thinking, it's one life and what do I with it? Now, there might be other people, who, in their journey, have chanced upon the authentic life that they want and kudos to them. But most of us are not there. So now we have to work through this journey. My advice is to give yourself a chance to go within.

S: Excellent. Thank you very much, Ram. It's been a very insightful and interesting discussion.

R: Well thank you so much for the opportunity to share my journey, and hopefully we can connect, you know, get other people to think along the same lines, and experience the kind of openness and learnings that we have experienced, yes?

S: Yes, absolutely, that would be fantastic.

Key Learnings:

- When you are really clear on what you want, you will attract it.
- When an opportunity presents itself, be ready, and grab it.

- Sometimes what you need to learn will show you what you can offer others.
- You may need to go within to shift the things that are happening around you or to break down your barriers.
- Don't work in a vacuum, get support.
- Finding your purpose and going on that journey is an opportunity to grow as a human being.
- When you're on the right path, synchronicity kicks in and provides you with some extra support. Watch out for synchronicity in your life.
- Important characteristics for people starting out on a new path are self-honesty, authenticity, connecting to your inner power.
- It's important to love yourself and know that you are enough.

You can find Ram at:
http://www.innr.me/

BODHI – BREATHE PROJECT

S: Thanks, Bodhi, for joining me. Do you want to start off with a bit of background about yourself?

B: Yes, what would you like to know?

S: How about what's happened for you over the last few years? What your life journey has been to bring you to where you're at now. And then it would be great if you could talk about the Breathe Project you've set up.

B: I spent a lot of my teenage years, from when I was ten until twenty-one, competing in surfing and not really looking after myself, not looking after my body. I was brought up vegetarian, with great food, but I didn't really care about my body. Then, a few years ago, I was in Bali, and I injured my knee again on my first surf. On that trip, I was meant to catch up with some friends in Bali and I was excited, and then, my first surf and I tore the medial ligaments in my right knee. And that was the second time in about six months.

S: Ouch.

B: Yes. And I thought, I don't know what to do. You know, I'd just got there, and I had three weeks. I hadn't seen my mum for almost two years and she messaged me to say that she was in Bali too. She said, "I'm going to a retreat with Nicky Knoff and James Bryan from Knoff Yoga up in Cairns. It's going to be an awesome yoga retreat." I'd been getting into practicing sun salutes before I'd surf, for about the last six months, so it sounded quite exciting. I had no money, but Mum said she'd pay for half of it, so I ended up going to this retreat. It was at that retreat where I did the exact same breathing technique that we're teaching to the kids in schools now. I think it was the first full breath I'd ever taken. The first time I'd even ever thought about taking a full, deep breath. And I realized, wow, there's something there, you know. And not just in the breath, but these people that I was surrounded by were full of life. They were eating well, their bodies were in shape, they were glowing. And that was the start of it, for me. I thought, these guys get to teach this, be on retreat, get paid for it, eat amazing food, hang out with good people, and just share something that makes them feel good. And that's when I thought, yes, I'm going to go to India, I'm going to study yoga, and I'm going to be a yoga teacher.

It didn't happen straight away, but I started practising. I was practising each day, doing salutes. I think it was about eight months later that I went to India. I thought I was going to study postures, right, study asanas, because that's all I'd known, and all I'd done. Anyway, I went over there, and I had two teachers – a Japanese lady and an Indian guy. The main teacher, the Indian dude, whenever it came to asanas he would look out the window. He would just be talking about it, oh yes and shoulder stand and, oh this, right leg up. There was no instruction, there were very little cues, which I love now, but back then, I'm like, "Man, tell me how to do it, give me the exact things that I need and want to know." I wanted to know the names of everything and he didn't care. He didn't give a shit about the names of anything or how they were done. And yet, when it came to breath work, when it came to pranayama, he lit up. He was sharing wisdom, not just technique, and the way he presented it was incredible.

During the training, I started to heal from having sinusitis, hay fever. I started to use the neti pot and started to breathe and practice each day. It's amazing, I haven't had any of those health issues since that time, issues I'd had for my whole childhood, my whole life. Not one person in my life ever said that you could cure hay fever, you know. My mum's an acupuncturist, a naturopath, and my dad's an osteo-naturopath, and I don't think they even knew that you could cure hay fever. So I'd taken natural medicine for it, things they would concoct. I'd had acupuncture for it, and nothing worked.

S: Wow. So, the hay fever was cured through the breath work?

B: Through breathing, yes. Through the practice, yes. It wasn't one thing, I'm sure, but through looking after myself, through being more of myself, and through using my nose to breathe. I had been a mouth breather. I couldn't even remember ever taking a deep breath through my nose. And, at the start, when I was using a neti pot, it was so blocked I couldn't even breathe through my nose.

S: And, sorry, what's a neti pot?

B: A neti pot, so it's a little pot, you may have seen them. You put warm water and sea salt in it, about a teaspoon of sea salt. You make it as salty as the sea, then you tilt your head and put the little nozzle up your nasal passage, and then it pours the water through the passage. And, then you do the other side. Yes, it's amazing clearing for the sinuses.

S: So tell me a bit about the Breathe Project and how on earth you managed to get that started? I'm a big fan of doing stuff in schools that creates change. I think what you're doing is amazing, and I just would love to hear about how that started and how you managed to get it going. You seem to have got a lot of momentum happening in a very short space of time.

B: Yes. Thank you. It has been amazing. I've always wanted to work in schools. When I left school, I wanted to be a school teacher, but I wasn't a fan of the school system. And so it feels funny now, walking in there, in the schools, but not being a part of the system. It feels incredible, actually.

The Breathe Project began when I was sitting [meditating]. I'd been feeling a bit lost, and I'd just finished up working as a psychosomatic therapist in Brisbane, at Soul Space. I had enjoyed it, but there was something missing, and I'd been feeling that I was a little bit out of integrity. I was a bit anxious; I was going through the motions of this is who I am, this is what I am doing. So, I decided to sit at a Vipassana ten-day silent retreat meditation, up at Pomona.

Anyway, I'm there on the second night, and I'm sitting there, and I get this visualization of Buddha. He dropped down, and he said, what are you doing here? And he just said those words, repeated them a few times, and then disappeared. And I was thinking, what was that all about, you know?

So, that was the second day. And, of course, on the first day I wanted to leave, on the second day I wanted to leave. On the third day, I'm like, get me out of here, but I knew that after he said that, I knew I'd committed to the ten days, so it was like another test for me. [Anyone who's done a ten-day Vipassana retreat will know this feeling.]

And on the fifth day, I'm sitting there in meditation, probably thinking about something, and suddenly I get this visualization. The first part of it is myself with a microphone in front of five hundred kids at school. And Kat, my partner, she's doing a handstand next to me in her funky acro tights. And behind us is a sign saying Breathe Project, in blue writing, and under the sign it says, for the community from the community.

And I was like, this is interesting, wow. And as soon as I saw that image, I got flash imagery of how the whole project was going to work. So, the colours, the people, the website, the formula to make it work. Everything came to me in that moment; how I could help kids in school on a larger scale, and create change, and keep it sustainable and allow it to have longevity.

It all just dropped in. I saw flashing imagery in front of my eyes, it just went tick, tick, tick, tick, tick, for a long time. I was thinking, at the time, about ten years ahead. I saw this vision, and then I saw flash imagery of my life, from when I was very young, and Kat's life, my partner. All the way from when we were young until the point when we met, and why it was us, and why we were starting and creating Breathe Project, and why we were going to share the wisdom of breath.

So that confirmed it. And then it was like, we're doing this. And, my partner was in the Himalayas at the time, she was trekking over there. She was in her second month of a two-month trip. And I couldn't tell her. I'd had this amazing idea and I couldn't tell anyone. I'm sitting for another five days in Vipassana thinking, get me out of here, I've got to write these ideas down.

S: Yes. I was going to say, that could be torture, having that insight right in the middle of your retreat.

B: I know. I was trying to let it go, and to think, it'll happen if it's meant to happen. And then I came out of retreat and I didn't tell anyone. It was the middle of winter, and I remember thinking, I've told so many people ideas in the past and had unhelpful feedback. I'm going to incubate this, just let it incubate over winter, and then I'll bring it out. So I only told two friends of mine, and one friend's like, man you can't tell Kat on your next phone call, you haven't talked to her in ages, you'll overwhelm her. So, then Kat calls me, two weeks later, she gets some phone service, and I'm talking and it's, like, surface talk, you know, ah the weather and it's cold over here. And I'm keeping it really, like, mellow. And then I said, I've got to tell you something, and I told her the whole Breathe Project spiel. And she was like, oh that's funny, I wrote something similar to that three days ago. And, I'm like, you're kidding me.

Kat and I met at an Art of Breath workshop that I was teaching at a YogaFest on the Sunshine Coast. I saw that this girl, her name was Kat, had 'breathe' tattooed on her right ankle, and

I wanted to know the story behind it. And Kat said, "Because it was there when no one else was." Of course, we connected, it's one of those crazy things that's just meant to happen, I think.

S: That's a cool story. Now, one of the things that stops people doing amazing things like what you guys are doing is the financial aspect of moving into a new area. So did you think, I really want to do this but I've got no money, or did you just think, I'm going for this and the money doesn't matter?

B: Yes, I run on passion. The money just shows up, it needs to, to support it, you know. To me, it's all about the commitment. So to get the Breathe Project going, we hid out for two weeks, and we wrote everything down. We didn't want to take any advice, because any advice, you know, it dilutes our rawness, it dilutes our potency. And, we start to think about the business side of things or what needs to happen for this to progress. All the other things that come from people's past experience. From their fears or what they've learnt. We're all going to learn our own way. So we hid out, we wrote it all down, and then we thought about the money. We were going to crowdfund. And I knew that we'd earn enough money, we'd have enough people donate to fund it. I was hoping to get the whole year funded, but we didn't see that money flowing in. It was a bit like, oh man, damn, because we had a high expectation.

S: Yes. Which is good, that you had high expectations.

B: Yes. But, that expectation got a bit shattered. However, it helped us be more present with it. More focused on where we were at, and aligning it more to right now. We put out there the dream of a year of doing this in schools and wanting it all to be funded. We raised $13,500. And, when I tell people what we've done with that, they're like, are you serious, with only $13,500? We've reached sixty schools with that, we've built up resources, and we've built a website.

S: That's amazing. You've been to sixty schools already?

B: Yes. Sixty schools. And, most of those have been whole

school approaches. So we've worked, during one day, on three workshops with the youth. We see every single kid in that school, high school or primary school. And then we work with the teachers for half an hour after school, in the staff meeting, all together, on integration. With the teachers, the focus is first on self-care and to help them out during stressful times. And then, for the teachers to be leaders, examples in front of the classroom, to be able to be the role models they want to be.

S: That's amazing. And how did you, if you don't mind me asking, how did you get into the schools? Or how did you decide which schools to approach? Have you gone through a hub to get to the schools? How has that worked?

B: This part's funny. I'd say how it's worked is from being very naïve, or ignorant, yes…

S: I like that things have worked through a naïve approach, and I like that you haven't taken advice, because it's almost like you're stepping outside of normal social constructs. And, sometimes being naïve and not getting advice from people can help that. You know, you do things differently, don't you?

B: Yes, well, otherwise I'm not being myself and then I feel trapped. Or I feel I must do it a certain way because that's just how it works and then it's like I'm back at school getting told what to do again.

So, being really naïve, we made it happen. We knew it was bigger than us, and we knew it would take more than us. Our golden rule from the start was, we're not going to approach one school, because we want it to be an empowerment process, even before we get there, so then they're welcoming. So, how it works is a parent, a teacher, a student, or a local can register their school online and express interest. They put in the school's details, their details, the school contact name, and the school contact number, whether they're a parent, teacher, principal, or a student, and then, from there, our info pack gets sent directly to them. And it says, please print this out and hand it directly to your school contact, and make sure they get in touch with us within the next three days to book their timeslot, to book their day. So it's very simple; I think it was three steps. And then schools started calling us, and I think we had fifty schools registered before term one started.

S: Wow, that's amazing.

B: Yes, it's been an amazing approach that I didn't have a clue would work. But there was a bit of hope in there, and a bit of trust, because otherwise it was like, there was no plan B. There was not a plan B of cold-calling schools, because that just wasn't going to happen. Sometimes four parents from one school have registered, and they've all taken it in to the school, and the principal has said to us, we've had four parents come to us with your information, so we really want you to come to our school. It's better that it's coming from them.

S: It shows there's a real need.

B: Yes. And because it's coming from them, that automatically brings openness, a willingness for the change. Not just for us to come in there, but a change of their approach at school as well, you know, and to put wellbeing right up there. So then the kids are more open too.

The only school where we didn't use that approach was Kat's old school, Maffra High School. She called them because she wanted to go there on our east coast trip, down in Victoria. It was really, really interesting because there was no openness at all from the kids, from the teachers. They hand-chose students from the school, the ones that they thought wouldn't play up, to be in our workshop. We're like, are you kidding me, bring everyone, you know, it's for everyone. So, it was an interesting experience, and it took us back to our golden rule.

S: Good to have the contrast, to see the difference between the approaches.

B: Yes, so that's continuing – right now, as we speak, there are schools registering daily.

S: Yes. And are you still crowdfunding?

B: No, we've transformed. The crowdfunding covered our whole east coast trip, which included the exposure to our sixty schools, and resources for those schools. We've left them with classroom posters, little Breathe cards with our cartoon character, and some pocketbooks for all the kids in class to use that have facts about breathing and our technique and how to connect with it daily. And now we've shifted our whole model, which was what we were going to do from the start, into schools paying for it.

S: And now schools still register with you on the website, but they need to pay for the sessions?

B: All the same, yes. And, out of sixty schools we had, I would say, eighty to ninety per cent didn't know that we were doing it for free. They asked us, at the end of the conversation, after they had booked in, they'd ask Kat because she's the coordinator, how much is it. And she'd say, well, actually, it's free. And, it's not free in the way of a charity, it's free in the way of people, real people, human beings like me and you that work whatever jobs, and do whatever, have paid for this. You know, it was like prepaid, you know?

S: Yes. That's incredible.

B: Yes. So it was cool to make them aware of that too, that it's not just us going out there saying, we're going to do this for free and peace and love and positivity, you know. No, this is prepaid, and that's how much people value, not just breathing, but change.

S: Well, I can't imagine the impact you've had already. I found it interesting that I was told what you were up to, and then a couple of days later, at my work, I saw one of your Breathe cards. And I thought to myself, the word's spreading, people are getting this great information.

B: Yes, a lot of people are using our little cards and 'Just Breathe' stickers on their computers. Right there at the bottom of the computer, if they're in the office, or having little reminders on the fridge, or stickers on the back of the phone. It fits the iPhone 6 perfectly or the iPhone 5. So it's a good reminder. What I've realized after doing this is every single one of us knows, but a lot of the time we just need a reminder to take a deep breath.

Because we know, and our body's natural response is to do it. So we're not teaching anything new, it's ancient and our body naturally does it. We're just bringing the awareness back to it, and trying to integrate that awareness, or help people integrate their own awareness more and more and more on a day-to-day basis.

S: Yes, it's fantastic, so inspiring. What would you say to people about how you feel now? Do you feel like you're living on purpose? And, I guess, what are the benefits of living on purpose, and how would you feel if you weren't?

B: Yes. How I'd feel if I wasn't doing it, I wouldn't know. But if I went to my past experiences of working in, you know, restaurants, working in the kitchen, it's a feeling of being trapped. It's a feeling of your anxiety rising through the roof. Sleepless nights, and why am I waking up, you know, what for? Yes, and now passion and purpose are number one, because it's about becoming more of yourself, and that feels liberating for me. That to me is freedom, you know. Health rises up, the physical body rises up, to meet that. The Breathe Project is amazing, and people ask, why did we do it on the coast, and I say, because we love living at the coast.

For us to turn up in front of a whole school of kids and teachers, day in and day out, and do four workshops a day, we've got to be in the best place we can possibly be in. And that, to me, is focusing and prioritizing the things that make me come alive. And that's how my last few years have been.

S: It's awesome. And I love how you mentioned earlier that the Breathe Project is bigger than you guys. Do you feel like that's part of it, having something to focus on that is bigger than you?

B: I've got a quote, and I've got to read this quote for you right now. It says, "When you have a purpose greater than you, it frees you up to be the best you can be. It's no longer about ego, you give yourself permission to be as successful as you possibly can because you know your purpose is greater than that. It empowers you."

S: Very nice.

B: I love that. And, when I first discovered that I was on such an ego trip. We're all on ego trips constantly, but I was on such an ego trip and my business was Bodhi Whitaker. But I was like, I couldn't see beyond just me, you know, really. And I saw that quote, and I'm like, I've got to write that down, man, I've got to, I want to, yes, it's bigger than us. But, no matter what, it's always bigger than us. If we're working in a small, little corner just doing our little thing in a little café, it's always beyond us, you know.

S: Yes. I think that's what I find inspiring about my work. A lot of what I do is about helping people to move beyond themselves. And, I love that I'm finding more and more people like me who are living their passion and aren't stuck in this, you know, have to go to work in a nine-to-five job, and feeling trapped, and all that sort of stuff.

B: Yes, because it's been made up.

S: Yes, exactly, and it doesn't have to be that way. But sometimes it takes a bit for people to see.

B: And that's the journey. I continue to find with every step, it doesn't get easier, it takes more, it takes all of you. To me, it's easier to work for someone doing something mundane, or just something okay that you can do, but it's soul-sucking, you know.

S: Yes, you're right, and it's more challenging to do something that's living your purpose, but it's more fulfilling, isn't it, on the other side?

B: Exactly, yes. But you're going to face, when you're doing that, so much more of your deeper self in every moment. It's like you're more vulnerable in every moment, you know, you're more sensitive and receptive.

S: And do you find that means there's more transformation going on as well?

B: One hundred per cent, yes. One hundred per cent growth on all levels, constantly, you know. I don't think you can be doing something that you love or doing something that feels so purposeful to you and not be growing.

S: So if anyone reading this wanted to donate to your project is that still possible?

B: Yes, one hundred per cent yes.

S: How would they go about doing that?

B: On our website, http://www.breatheproject.com.au/, you can find a little link up the top that says donate.

S: Awesome. Very good. I really like the design of your brand. It's funny, because my background's advertising, and I'm really picky about stuff like that. When I first saw it I thought, this is really professional but it's also very authentic and unique. All right, well we might finish up. Thank you very much for your time.

B: Thank you.

Key Learnings:

- Opportunities can arise, sometimes, when you least expect them. Be open to new ideas and opportunities when they present themselves.
- Sometimes it is better to follow your own vision and not let things get diluted by the worries and concerns of others.
- Do what you think is the right thing to do. Remember that advice from others carries with it their own biases and past experiences. Sometimes your naivety or inexperience can be what makes you do things differently, and may even bring you success.
- I've included the quote that motivates Bodhi: "When you have a purpose greater than you, it frees you up to

be the best you can be. It's no longer about ego, you give yourself permission to be as successful as you possibly can because you know your purpose is greater than that. It empowers you."

- Living on purpose can be more challenging than not. Not living on purpose can be soul-destroying. However, a life on purpose is likely to bring one hundred per cent more transformation.

You can find Bodhi at:
http://www.breatheproject.com.au

KATRINA – HEALER

The healer I interviewed would like to remain anonymous. Katrina is not her real name.

S: Katrina is a palliative care nurse who is going to tell us her story about how she found her passion. She'll also give us a bit of her background, and then we'll do some questions and answers. So, welcome.

K: Thank you. I was born in East Germany and came to Australia about eleven years ago. Because I was born in East Germany, I haven't always felt very free. Growing up, I couldn't leave the country, my parents couldn't leave either, so there was a lot of restriction in my life at the beginning. When I was eleven years old the wall came down in Germany, and then we had the opportunity to move into West Germany and to start travelling and to see more of the world.

S: So you were there when the wall came down.

K: Yes.

S: Do you remember, I understand you were quite young, the transition from when you were in the eastern part and possibly quite stuck with not a lot of freedom, to more access to the west and more freedom.

K: Yes, definitely. When I grew up we did not have certain things in shops. We weren't allowed to speak in groups on the street, everything was very controlled. I know that certain things weren't allowed, like watching certain TV shows, having certain magazines that are from the western side. So I was quite aware that I wasn't allowed to do certain things. My parents, probably more so, because they were older. I had family over in West Germany, and my grandad lived in West Berlin. They came to visit us sometimes, and my grandad, he always used to say, oh when I go home I'm going to put you in my suitcase and take you with me.

S: Oh, that's lovely.

K: And every time he said that, I waited for him to do it. But yes, when the wall came down everybody was very excited. We went over to West Berlin the next day, and everything smelled so good, and everything was just very different to what I'd known. The shelves were full in the supermarkets, everything was lit up at night, there was lots of advertising, all these things were very different. So then I moved to West Germany.

As for my passion, my career, I always knew that I wanted to be a nurse. I think I first mentioned this to my parents when I was about five years old. It was something I just knew. Initially it was midwifery, and then it moved to nursing. So I started looking for nursing jobs once I finished school, started my study in nursing, and was always drawn to the palliative care field. It was just something I was interested in. It might sound a bit weird, but I was interested in death itself. So when people are dying and what happens to them was always something I was interested in. I worked as a nurse for two and half years in Germany, in the oncological ward helping people with cancer There was lots of palliative care required. Then I had the idea, with a friend, to go to Australia and do some travelling.

I still had my job on the ward, so I took some leave, unpaid leave for one year, and the plan was to just travel. So off we went to Australia and started travelling, and after about six months my friend and I went our separate ways. She moved on with someone else, and I went to Melbourne.

Also, from when I was about eighteen years old I was interested in astrology and intuition. I've always felt people, from when I was very little. I could feel how people were feeling, and how I could make them feel better. It was just something that I've always had, and I felt that in German society it was something that wasn't accepted or encouraged. I've always felt a bit different, and when I came to Australia I felt like I could just be me, and I didn't have to explain anything. Either people just didn't care what I did or they were just okay with who I was.

S: So there was no judgement about what you were doing.

K: And, it was a feeling of, I'm free.

S: When you say free, do you mean free to do what felt right for you rather than feeling like you had to do nursing and stay in the palliative care area?

K: Yes, I didn't have to focus on just working and earning money and paying my bills. I felt like I could work on my own terms a little bit more, and continue my interest in astrology and energy healing. And after six months in Melbourne, it all just came together. I met someone who gave me a reading that said this was the starting point of my journey, and she said it was very important for me to stay in Australia. I already wanted that, but I think I just needed someone to tell me it was okay. So from there I needed to get my registration as a nurse to be able to stay in Australia. I had to get a new visa, and I had to do the English test. I felt very determined to do all of that, even though I didn't have a lot of time. And I had to ring my parents and say I'm not coming home.

S: How were they with that decision?

K: Of course, they said that's okay, you need to do what you need to do. At the same time, my mum didn't speak much on the phone to me because I think it hurt her that I wasn't coming back.

And it wasn't easy to not feel guilty about staying in Australia and leaving Germany.

S: It's quite interesting that sometimes it seems you need to have a complete change of environment to be able to make real changes in your life. Maybe shifting from where you were to Australia gave you the space to do what you needed to do. It seems you're able to leave some things behind when you move such a distance.

K: Definitely, I had different opportunities in Australia too. So the lady that I met who gave me a reading, she started offering courses to me. Things like intuitive development and energy healing, and I felt very drawn to those things. I would never have had this type of opportunity in Germany, because Germany, at that time, wasn't developed in that area. I felt like I was finding something that was more me, and it was very different to what other people do.

I didn't have to fit into society anymore. This is what I wanted, I wanted the freedom to be who I needed to be for myself rather than for other people. If I'd gone back to Germany, I would've been the same old Katrina to everyone. I wouldn't have been able to be myself for myself. I needed to be able to express that intuitive part of me that was always there from when I was a child and to do something with it. A lot of the time when you have something that's a little bit different, you need to have courage to use it.

S: So you were still in Melbourne?

K: Yes, I lived in Melbourne for about eight years. During that time, I had my first patient die in my arms.

S: Yes, being a palliative care nurse must be very challenging, you're dealing with people who are dying, and you're dealing with a lot of suffering. How did you manage that? You've said it's just been a natural thing for you.

K: I think death is, in my mind, another journey, it's not final. I think it's a privilege to be with someone in those last stages of life, and to witness that part of their journey with them, because it's very special, like a birth. You know, when someone gets born it's such a big thing and it's a celebration. It's almost the same with death, it's not a celebration, but it's a very important event, and it's something that everybody needs to deal with at some stage. I think it's a privilege, it's a privilege to be able to help the family and share with them some of the knowledge that we have as palliative care nurses. To give them care and love. It's a much more holistic approach to nursing rather than working on a surgical ward or a medical ward. So, working at a Hospice and Palliative Care Service, you work with the patients and the family very closely. Of course, you get attached to them at times, but I think that you've got to learn to set boundaries, and you've got to learn how to give deeply from yourself without getting too attached.

However, I think if I didn't feel anything anymore I would be very worried. When someone dies I always feel it, and I always have tears in my eyes when I see the family cry, and if I still have that I know that I haven't completely switched off. So I still want to feel, but yes, I do at the same time want to go home and live my life. It's been a learning process, and I think that's very important in any kind of work when you have close relationships – you also need to be able to develop your boundaries.

S: Let's go back to when you said that your first patient died in your arms.

K: When it happened, I was quite surprised. The nurse, my mentor, she said, go to that lady and help her have a wash in bed, and she was obviously ill but she wasn't at end of life. I helped her to wash herself and then, all of a sudden, she couldn't breathe, and I held her up and she just died right then.

S: Wow.

K: I pressed the emergency button, of course, just before that, but by the time the others came it was too late. I had quite a few of these events in the early stages of my nursing career. I felt privileged. I felt like at least someone was there for them, that at least she didn't just die in bed without anyone knowing about it or being there for her.

S: That's lovely. I like the comparison you made between birth and death. Isn't it interesting how much we celebrate birth, and yet we don't pay anywhere near as much attention to death, when it's another very important event. It's an important time, and yet we're so scared of death that we'd rather avoid it.

K: Yes, it's good to think about these things I think, even in our own spiritual practice, to think about the impermanence of life, of everything. In palliative care nursing you see it every day, that nothing is permanent, nothing will stay the same. Everything changes, and it's a great way of learning that lesson.

S: Yes, I can imagine. So going back to your journey. You were in Germany, you were there when the wall came down, you came to Australia, travelled, decided to stay in Melbourne, and then what was the next stage for you?

K: Well, I did quite a bit of study in terms of healing, and did my own practice, my own meditation practice. I went more and more into energy healing, and started doing healings on myself, and doing meditation almost every day. All this practice gave me a sense of peace.

Queensland had always been on my mind, and I always knew intuitively that that was where I was going to move to. It was about four years ago that things just fell into place for me. The place where I lived got flooded, so I got a notice to vacate because the owner wanted to sell. It was then that I decided I'm not going to move within Melbourne, I'm going to move somewhere else and I thought of Queensland again. I didn't know much about it, so I looked on a map, and I thought Maroochydore sounds good. I booked a flight to Brisbane, hired a car, just for a week to look around, and drove up the coast.

When I drove into Mooloolaba this song come on the radio, the words were something like 'Oh, I've got a good feeling', and that song never left me the whole week. I saw it as a sign that I was in the right place now. So I flew back to Melbourne, and started looking for a place to share in Mooloolaba.

S: Wow, that's quite brave. You just decided to move and then did it.

K: Yes, I have great friends in Melbourne, but I felt that moving was the right thing to do. When I got to Mooloolaba I met the right people, got my own place, and started building my business in soul regression, energy healing, and intuitive work, and started working at the hospice.

S: So you moved with no job to go to. You moved first, and then worried about what you were going to do later.

K: It just felt right, that's the only way I can describe it. I wanted to work at a hospice, and so I found a job, applied, and got it.

S: You are lucky compared to a lot of people, because you knew from a young age that you wanted to be a nurse. You followed that path, and you're still doing that job. And then, you've also always known that you're a bit different, that you've got this intuitive skill to keep developing. You've got two paths that you're following. You can do nursing, and then have your own business, it keeps it interesting.

K: I was very lucky I think. I would not have been able to stay in Australia if I hadn't been a nurse. Nurses were very much needed in Australia when I arrived, and that was the only way for me to get a sponsorship visa and to finally get permanent residency. So mostly it all lined up, but in between there were also periods in my life where things were very difficult too.

To be able to go with your gut feeling you've got to have courage, you've got to trust, and if you don't trust yourself then that's where the problem is. Trusting yourself is very important, and I think it's important to have something we can retreat into to get to know ourselves a lot better – something like meditation.

S: Interesting. So I've got a few questions. What are the main barriers or challenges that you've had to overcome?

K: My confidence. Believing in myself. Yes, that was a big one. I think a lot of the time I've put myself down because of childhood experiences and other things. The trusting, I think trusting myself.

S: So, for you, many challenges have been internal more than external?

K: Yes. I do believe that everything is a reflection of our inner world. So if we are peaceful inside then that's going to reflect out into the environment. I think my confidence was always a big struggle, but I can say that a lot of that has shifted, and I feel a lot stronger within myself, I feel more sure of myself now. Self-doubt is a good thing in some ways because we need to question ourselves, we need to always be on our toes.

S: What have been the benefits of following your dreams? What I've heard from you is that you've been constantly developing and evolving, and the more you found peace and contentment within, the more that's reflected into your outside world. Did you want to add anything to that?

K: I believe that striving for inner peace, doing the things that we love, and taking the time for those things is important. I love gardening, so I garden when I can. I love meditating, so I meditate. It's very important to just be with ourselves sometimes, because so often we get distracted. We are always with people, we are always doing something. It's very important to self-reflect. I think that's part of achieving your dream, because that self-reflection will give you the insight into your resources, your own world, your own dreams, and that's where you can draw your strength from.

S: There are so many people who never spend a minute alone, they're always on their phone, or their computer, or they're talking to someone, or they come home and turn on the TV. I wonder, when are they getting any down time? It is these people that will quite often say, oh, I'm so busy and I'm so stressed. And I think, well, maybe if you spent some time within rather than looking at what's happening outside you'd feel more peaceful.

K: Yes, and that can be, you know, anything like an exercise, it can be meditation, it can be...

S: A walk on the beach.

K: Yes, it can be gardening, it can be anything. But just to have a little bit of time with yourself can make a big difference. It's self-development, basically.

S: What sort of traits do you think are important to ensure you're following your dreams? I guess that self-reflection and having some time alone are some of the things that you consider to be quite important. Is there anything else you consider to be important for people to be aware of?

K: Self-reflection and self-awareness are the most important things, because they allow you to make decisions based on what you want. So be patient and put the work first into self-development, and then make changes, instead of expecting everything to change immediately.

S: I think you're right. You need to do the work inside first, and maybe that's why people do get stuck. And, if you do just make some random decisions and it's not really coming from within and it's not part of a natural flow, it doesn't work. There's a tendency for us to think we want a quick fix, and we can change things straight away, but that's not how it works.

K: Take some time for yourself and consider – who am I? What are my talents, what are my strengths? What am I good at? Take small steps, make changes within, and then everything will flow from there.

S: One of the things I'm quite interested in is how you manage your time and what your support network looks like.

K: I think it's very important if we've had a stressful day at work, that we come home and just have that half an hour or an hour of down time. Have a bit of a chat with your partner, or someone else, or do something that relaxes you.

I think it's very important to surround yourself with people that support you. Of course, we all need people that are close to us, but it's not the quantity of people, it's more the quality that's important. And being able to be who you are without being judged, and to be encouraged. Encouragement is very important, and support.

S: Yes, it's a bit of a generalization, but I do feel that society tends to want to pull people down rather than push people up. So I've found that it's important to surround myself with people that are very positive. Not in a superficial way, but in a very supportive, helpful way, to allow you to do what you need to do.

One of the myths that I'm busting is around following your dreams and not making money. Generally, the people that are following their dreams aren't focused on money, and I think we can get a bit obsessed with money, but I'd like to hear your perspective on this area.

K: I think it depends on the circumstances of the person, and it comes back to flow, like we've spoken about before. If we're able to find that flow, then I think we are always provided for. Personally, I can only speak about myself, I don't need heaps of money to be happy. I do like to be able to go the shops and just buy a T-shirt or a dress or whatever if I want to. So yes, of course, you know, for me personally I like to have that little bit of luxury.

However, I think it does come back to being in the flow and I think it does come back to living on purpose. If we are able to do that then everything else will be provided. I don't know how to answer that question because I think it depends on the expectation that you have, how much you can live with, and how much you can't live with. You know, if you have a mortgage to pay, of course, you have a lot more pressure. I do generally feel that what is important is living your purpose and living in the flow and doing as much as you can that is in alignment with yourself and what you feel is good.

S: Yes, so it depends on how much you are willing to simplify your life as well.

K: Exactly, yes.

S: I feel so much more content now. It's funny, because I'm living a lot more simply than I used to, and yet I still feel really rich. I live in a beautiful apartment, I can buy the food I want, and I can do the things I want to do. And, now I have so much more time for myself and that's invaluable.

K: Yes, I feel exactly the same way. I probably have less today than I had two years ago, in terms of money, but I feel much richer and I feel like I have more time, I have more time for myself. And I think it's a trade-off, you have more time for yourself if you work less, if you have less money. It gives you more time to self-reflect and work on yourself.

S: That's another thing that we don't value enough – time.

K: Yes, and the peace that can come with that.

S: Now, if I think of a nine-to-five job it makes me cringe. The whole idea of having to be in an office nine to five, five days a week, and I mean, it's not nine to five for most people, it's eight to six or whatever it is, and it's not good. That's where I think you can get a bit caught up in that whole money cycle or an ego money cycle.

K: Yes, so it depends on what you want for yourself. How much you need. But, we don't need much to be happy, that's for sure.

S: Yes, you're very right. So just a couple more questions now. Looking back, is there anything that you might've done differently on your journey?

K: I think it's all just a learning journey, and everything is a lesson. If we step off the path, then we've learnt something, and we can come back onto it. I can't say I've got any regrets. What I would have done differently, however, is I would have like to have believed in myself more. I think sometimes that's held me back, things might have gone quicker if I believed in myself more. I think that might be the only thing.

S: Yes, a couple of other people have made that same point. I guess we develop in our own time. So, just before I wrap up, is there anything else, any advice you'd like to give people who are trying to following their dreams, anything else you'd like to add?

K: Take your time, be patient, and take small steps, and start reflecting. Take the time to reflect on yourself and look at who you are. You know, what your strengths are, what you like doing, and assessing the important things in your life, and go from there. But don't put too much pressure on yourself. Don't think everything is going to change tomorrow, because it won't and that will be frustrating. So strive for peace rather than going for a certain goal. I mean goals are always good, but I think overall, peace is probably a very good goal and everything else comes from there.

S: So, to summarize a couple of key points. Firstly, you mentioned at the start of our conversation that when you moved to Australia you felt you could do what you wanted to do rather than feel you had to meet society's expectations. That story struck a chord with me, because I think that's where we can get stuck – doing what everyone else expects us to do and not listening to ourselves. Secondly, you discussed how important it is to work on ourselves, to look after ourselves, and to trust ourselves. Those are two great points, so thank you very much.

K: No, thank you.

Key Learnings:

- Sometimes you need a complete change of
  environment to make big changes in your life. A
  change in environment can help you to move beyond

your past experiences and shift beyond others' perceptions of you.

- Take the leap – go where you feel drawn and opportunities will present themselves.
- You need to learn to trust yourself, to believe in yourself. Have confidence in yourself. While self-doubt has benefits, don't let it rule you.
- Are you compromising what's best for you by conforming to what society or other people expect you to do or be?
- It's important to constantly develop and evolve, and as you develop peace and contentment internally, that will reflect in your external world.
- Do the things you love. Take time for you – whether it be gardening, a walk on the beach, meditating. Make time for it.
- Take time for self-reflection. Have alone time. This is part of achieving your dream. Self-reflection can give you insight into your resources, your world, and your dreams, and it can help you identify where to draw strength from. Think about trying meditation to assist this process.
- Strive for peace.
- Be patient with change – make sure it's what you want, and that you are ready, before making dramatic changes.
- Take your time, be patient, take small steps and start reflecting.

CHRISTINE – REIKI TEACHER

S: I'd like to introduce you to Christine Maudy, originally from France. We're going to talk a bit about her background and how she came to practice and teach Reiki.

C: Okay, it's a bit complicated. However, I think I'm now at a stage where everything makes sense, because I believe that everything we learn has a reason and a purpose in the big scheme of our life.

I began by studying psychology. I left my family and went to university in Paris. I had to pay for my studies myself and for my rent. Luckily I met some people who offered me work in a centre that was looking after children with social problems. Most of these children didn't have families, or they had parents in jail, or they came from some other difficult circumstance. I went to the centre to sleep with the children at night, a couple of times a week and sometimes on the weekend, and that was enough to pay for my studies and for my rent.

The woman in charge of the centre liked me; I think she liked my energy, and she could see I loved the children, and she knew I was studying psychology. One day she said to me, if you want to, come and work full-time and I will pay the school fees for you to study specialized education. I agreed, and did this program for four years. It was perfect as I enjoyed working with children and learning. When I finished the course, I had a Diploma in Special Education. I kept working in this area for a few years, working with children suffering from autism, physical disabilities, and many other issues. However, I found it too challenging – I was only twenty years old.

S: That's quite young.

C: There were many children who couldn't walk, who were bedridden. It was mentally and physically quite hard and eventually I knew I couldn't keep doing this work. It was at this time that I started doing some fashion designing on the side. I was making all my clothing, everything from scratch. I was drawing the patterns, I was doing everything, and I was also painting on timber at the time. Then I met some friends who wanted to open a small gallery in this place near Paris where I used to live.

We were three couples taking turns to work in this place. On the days I wasn't working I was selling some of the clothes I was making, some of my paintings, painting to order, and doing commission work for people. It was good fun. Then my boyfriend said to me, all your girlfriends are wearing what you're creating, so why don't you go for it in fashion? And, I thought, okay, why not? So, I put all my money, all my savings, into hiring a stall at the big fashion show in Paris, which was very impulsive and pretty crazy.

S: Wow, very daring.

C: Yes, it was quite daring. But I didn't have any doubts, and I had to try, because if you don't try, you don't know, so I had to be brave. I put all my money into, not only having a stall, but building a full fashion collection. I drew the whole collection. I bought fabrics in Germany where there was a big show for fabrics, and I created a collection. I did the fashion show with my boyfriend, it was an intense four or five days.

And it worked. The retailers loved what I was doing, and we

filled up with orders. And, at this stage, I did not have a clue how I would be able to make the quantities requested, how I would be able to deliver what they wanted. I just went to the show with a beautiful collection to test, and then we got all these orders. I was very excited. So when the show was finished, I immediately started to call people in Paris who could do all the sewing and other requirements. I was very lucky because I met this guy who had people working for him, and they were doing big quantities for fashion designers. I started to work with him, and then the orders became too big, so I had to find a little factory in Brittany. Then it became more complicated because you had to plan things out a lot more.

I did this for quite a while, and after the third or fourth collection – you have two collections a year – I started to doubt what I was doing. I was not having so much fun. Drawing the collection, choosing the fabrics, yes, that was fabulous and making me happy. But sometimes people, instead of buying the dress and the little jacket that went with it, would say we just want the dress and we want two thousand of them. And I thought, ah, that's boring, it would make money, but it's boring.

So, I decided I would slow down and maybe look at doing something else. Then I just decided to stop the business and take a break, and the break was to have a baby, so that was a good reason. But the break didn't last long because, as a fashion designer with my own label, I had been interviewed a lot. I had been included in quite a lot of editorials, and I had made friends with many lovely journalists in all the best fashion magazines in Paris. One of my friends asked me, while you are taking care of yourself and waiting for the baby to come, why don't you help me to get some editorials for some of my products. And I said, oh yes, why not, that will be easy.

So I went to see a journalist I knew saying, well, look what I have, and I presented my friend's product. The journalist thought it was fabulous and was happy to promote it. I said to her, what should I do next? And she explained to me how to do a press release and the importance of having a good photo.

Anyway, I ended up being so successful in getting exposure for these products that some companies contacted me and asked me to do PR for them. And before I knew it, I had to create a PR and event management company. Then I started to do events and PR in all different sectors – food, tourism, beauty, etc. I loved it, and I did this for more than fifteen years with a lot of pleasure.

S: It's very interesting what you've said about your journey so far, because it doesn't feel like it's ever been about ambition or planning or anything like that. It seems like you did what you enjoyed, and then when you didn't enjoy it anymore you moved on to something else. You found whatever opportunity presented itself, and then ran with it.

C: Yes, when I look back, it's been about pleasure and happiness, what makes you happy, and I think that's very important, to do something that makes your heart sing.

It's not about having money. Of course, we need money to survive, but it's not good to go and work in a place that you feel is draining you of your energy. I have met people who were working in the mines because they could earn lots of money and very fast. But during this time they didn't see their partners or their families, and sometimes their families fell apart. And they were working in a very toxic environment. So money is one thing, but at what price?

S: It's great to hear you explain it like that, because one of the challenges I find when I talk to people that are frustrated in their career, is that they're very focused on money. And I always feel that until they can put that aside, they will remain quite stuck. Choosing a career is about pleasure and happiness and serving others.

C: And, you know, I think that when you're happy, when you're in your flow, everything starts working for you. And, even if you don't make, you know, a thousand million dollars, you make enough to keep doing it and to develop it. I believe that both work together.

But the first thing is to be happy. If you're not happy, if everything is just about earning money, life becomes very hard. And, sometimes when you talk about career, it can be about the ego. It's about what other people are going to think. Don't give a damn about what other people think.

S: So you helped kids, set up and ran a fashion business, and then set up a PR agency. And then, what after that?

C: Then it just happened that I was attracted to travel to Africa, to Sierra Leone. This inspiration came through my work. Part of my job was to buy all the newspapers and magazines and to collect the editorial content for my clients. So one winter weekend I was doing this work, and I saw in a French magazine this double page spread of an amazing photo of Sierra Leone. The picture was of a beautiful, pristine beach, and the editorial was saying that the country was just opening for tourism. That place really called me. And I said to my husband, that's where we are going on holiday. And he said, yes, why not?

When we arrived in Sierra Leone my life changed. I had a big spiritual awakening, I think. At first I didn't understand what had happened. The day we landed, I remember it was very warm, and we had left six hours earlier from Paris where it was freezing. Everything was overwhelming, the smell of the country, the atmosphere, the energy. I felt a tsunami of emotion, I felt completely excited. All my senses were completely alive. I couldn't sleep the first night, I was too excited.

The next morning, as soon as it was daylight, I went out and I thought, wow! There were mountains with the jungle behind us, these beautiful mountains, this beautiful yellow sand pristine beach that went for kilometres. It was just absolutely amazing. It was exhilarating and at the same time I felt like I was home in a way. It was peaceful and exciting at same time, which is an amazing combination.

I spent two weeks there with my daughter and my husband. The people who ran the resort had, like me, fallen in love with the place and decided to leave behind the life they knew. When we left, I felt very depressed. However, I went back to Paris. I remember the first day, I was still happy to meet my team because I had a beautiful team of girls working for me, and I adored my clients. I was always joking with people who, like my husband, were more money oriented, that I loved my clients so much that I would work for free. But, you know, my heart was not there anymore. Suddenly, there was something missing. It was very strange.

I woke up every morning almost crying and feeling that I needed to go back. There was something I needed to understand. And because I'm very stubborn, I went to this tiny little office in Paris where there was the only airline going to Sierra Leone at the time, and I also talked to the people running the resort there, and I said 'you need me.' You're just starting, you want to grow, and you want people to know about your place. I have been there and I just love it, I know the journalists, and I can bring them to your place. And they said, we don't have a budget but we can pay you with your flights and free holidays, and I said, okay, deal!

I did this for seven years. Each year I went three or four times a year taking journalists and visiting on a holiday with my family and it was magical. And I started to understand why I was so fascinated by this place – I think it was my connection with nature when I was there and with people living their passion.

S: You'd seen people who had uprooted themselves and followed their dreams.

C: Yes, they were following their dream. So I convinced my family that we needed to move because that was my dream. My dream was to live like the people that ran the resort. But in a different place, because I knew that Sierra Leone, or Africa, was not what I was looking for. I wanted a connection with nature, I wanted to live passionately, and I wanted to have new experiences. And that's how we came to Australia.

For our next holiday, we booked a trip to Australia. We arrived in Darwin, and loved it. We came three times to visit a lot of different parts of Australia and we were immediately convinced that this was the place. And everything ended up working perfectly. My husband was invited to do a trip to Australia with the French Chamber of Commerce. He came back even more convinced that it was a good place to move to.

So we sold everything. We sold our companies, we sold our big apartment, we sold everything. There was no going back. And we moved to Australia. It was like starting a new chapter. At one stage I considered starting again in PR, but then I thought, let's look for new inspiration, why not?

I had plenty of time and I started to paint and paint and paint and paint. After a few months, there were paintings everywhere. We had made quite a lot of friends on the Sunshine Coast and people were asking me, they're really nice, you know, can we buy your paintings? And then I thought, okay, I'm going to organize an exhibition. I met the people from the regional gallery, and had the first exhibition there. I sold thirty per cent of my work, which was amazing, it was very encouraging, so I then went full on. In ten years, I had twenty-seven solo exhibitions, and more than sixty group exhibitions. Not only in Australia but also overseas.

S: Wonderful. I'm curious, were there any challenges or barriers that you've had to overcome along the way? Or was it all smooth sailing?

C: Well, it's about learning. In fashion, you have fun creating something, and then people say they want a thousand of a dress. That's a big challenge, and a lot of people could say it's too complicated, I can't do that. Or they could spend a lot of time thinking about it, which I don't do usually, that's just not me – I work to find a solution.

With the paintings, I kept learning new techniques, I went

back to art school in Sydney during summer to work with some artists and teachers to learn some more tricks when I felt I needed to. Richard Branson, I think he said, if you have an amazing opportunity, you say yes, and then you learn the skill. Basically, without thinking, that's what I did.

S: It's almost like the word barrier or challenge doesn't exist in your mindset. When something comes up, you think how do I solve it, what new skill do I need to learn? And then what came next for you?

C: While I was painting, I'd been doing yoga and meditation. A friend introduced me to a place with amazing energy where people were doing meditation. People with health issues were going there to heal. I went there for quite a few weeks, maybe even a few months, and I met this woman who I became friends with. Then, one day, I learnt that she was a Reiki master and a Reiki teacher. So, I started to read about Reiki and it just made sense to me.

At first, it was for myself, for my personal development, and then to help my family with some health problems. I learnt Reiki level one, and I had a few treatments, and loved it. And then, of course, I did the level two and, decided to progress, because I love understanding, I love finishing things. And then we did the Reiki master. She was doing it in two parts – Reiki master for your development, and then for the people who wanted to teach Reiki. And I told her, well, I'm not doing the Reiki teacher, I don't think it's me, and I don't think I will teach. And she looked at me with a little smile, and I thought, what is that all about?

Just a little smile, which was a bit of irony, I suppose. I went home, and I felt amazing after learning the Reiki master. And of course, the next morning I got up early, sent her a text message, and she said, you're coming, I know. And so, yes, I did the teaching training and then I decided to start to do this for work. So I started to give treatments, and everything started to flow nicely, having people, then I decided that, yes, I will teach.

What I love about Reiki is that it's a method of empowerment. Reiki is not just another modality. Reiki is a complete system, not only of personal growth, but of healing. And, I love teaching Reiki, because I see people changing so much, and sometimes it only takes a weekend. You see people arriving a bit shy or with a physical problem, and after a weekend, they believe they can learn, they believe they can change, and they seem stronger.

S: I see Reiki as very similar to practicing meditation and mindfulness. When you really practice, it becomes part of you, and you just live it.

C: Exactly. It's a way of life.

S: One of the things I've noticed talking to different people who are living their purpose, is that they tend to be very busy, active people. So how do you fit everything in? Because, you've got a family, you've got your business, you're doing Reiki healing and you're teaching Reiki, you're the vice-president of Alliance Française, and you organize events as well. And there's probably other things you're doing that I don't even know about. So how do you manage all that? Because you're also very calm, you're not claiming to be overly busy, which says to me that you've got everything under control.

C: I have a full life, but I also have a fulfilling life. Because I'm free, I'm free to organize my work for when I want to do it. If I need to write an article for a magazine, or if I'm organizing an event, I can do it late at night. I just feel that everything I do gives me pleasure, makes me happy.

Everything I do, whatever it is, preparing food because I have people coming for a retreat or workshop, or writing something, or creating a new workshop, or a new course, learning, studying, because I want to keep learning and I am interested in many things, always give me pleasure.

S: So everything you do gives you energy?

C: Exactly.

S: I remember when I quit advertising and took a year off to travel, and one of my friends said to me, just notice what gives you energy and what takes energy away from you. And I think that's where a lot of people struggle with their work, because it's...

C: Draining.

S: Yes, it takes all their energy away from them. And it's true, when you're doing something that gives you pleasure, it gives you energy, and then you can do so much, can't you?

C: Exactly. Very true. And, you know, doing the accounting for my business was draining me because I had to force myself, and it's not good to force yourself. It's better to, you know, to give that kind of work to someone else and to spend that bit of extra money. But to be in your groove, to be in the flow, and to do something which is satisfying for you – that is the best.

S: I've noticed that you connect very well with people, and that you have very strong connections with a lot of people. That seems to be one of your skills that has made you so successful. And what about a support network? I think you've mentioned that you have a mentor?

C: Well, throughout my career my main support was my husband – we both helped each other, we were a creative team. But since we separated, I think, apart from my mentor, I have some great friends. I have five or six very close friends that I can count on, and vice versa. But for the work I do now, my main support is my mentor. She and I have a strong connection, and she always has enlightening ideas, and an amazing sense of humor. She's a psychologist. She learnt Reiki when she was younger, so she knows energy, and she works with meditation. She uses lots of different modalities, and she has taught me a lot.

S: And what traits do you think are important for following your dreams?

C: I'm stubborn and resilient. So when I want something, I don't change my mind after five minutes or because there is a little obstacle or big obstacle. I just keep going. Unless I realize that I don't like it and it was a mistake, then I recognize this and make a change.

I think I'm intelligent, a good listener, creative, and I love challenges. I love having fun and learning something new.

S: Those are great traits. I think those are all important factors for being able to follow your dream. And I think the most important thing I've seen is your connection with people. You're very good with people, and you have strong relationships.

C: Yes, I'm interested in people and I love seeing them evolving and changing, I love listening to them, and I love seeing people happy.

S: That's great. Sometimes, one of the things that stops people from following their dreams, is this whole money orientation, they believe it's not financially viable. I don't know what your financial situation is, but you've got a beautiful home in Woombye, and you seem to be doing well. Can you confirm that following your dreams is financially viable?

C: Yes. Well, we made quite a lot of money with our different businesses in France, and it was enough to decide to stop everything at the age of forty-five. From the divorce, I kept the house, which was good because it's the perfect property to do what I'm doing and it attracts a lot of people. These last few years I have been challenged on the money side. Living in a house that was built eighteen or nineteen years ago means renovations are required sometimes. And I have sometimes wondered, how am I going to pay for this? Then I just learned to trust. I had to work on another level of my personal confidence in avoiding fears around money. And, everything has always worked out fine. I think it's deciding to be in the flow, and not to be worried.

S: I wanted to mention that we're on the Sunshine Coast in Australia, which is a regional area. It's not a place where there are a lot of people with a lot of money. I think to be successful here is outstanding. So, looking back, is there anything you would have done differently?

C: The only thing, maybe, that I would change, is that I would have loved to learn Reiki earlier.

S: Those are all my questions. Is there anything else that you'd like to say, or any advice that you'd give others that want to follow their dreams?

C: Yes, I think it's most important to find something that makes you happy, and to be quite open-minded, to let go of your personal judgement. Because fear is something which blocks people, and worrying about what others think. And the fear of not making enough money. Sometimes we must jump into the void.

My advice would be, learn to trust. Let go of fear, and learn to trust. It is so important that you trust that there will always be a solution.

S: Yes, that's great advice. Thank you very much.

C: You're welcome. My pleasure.

Key Learnings:

- See challenges as opportunities to learn and grow.
- When problems arise, don't think about them too much, or let them stop you. Get stuck into finding solutions.
- Traits that are beneficial to following your passion are stubbornness, intelligence, love of learning, being a good listener, creativity, enjoying challenges, doing what you love, and having fun.
- Connect with people, create an authentic connection with others, and develop strong long-term relationships.
- Let go of fear and worry, and trust that the right solution will appear.

- Take a leap of faith – sometimes we must jump into the void for great things to happen.

You can find Christine at:
http://www.christinemaudy.com/
www.livinginspired.com.au

CYNTHIA – ZUMBA INSTRUCTOR

S: So, if you wouldn't mind just telling me a little bit of your background to start with.

C: Well, I'm from Chile, and I arrived in Australia in 2004, my background is in dancing. I used to dance at school, and then I danced at university. I used to do a lot of traditional dances, like South American dances. And then, in 2009, one of my friends took me to my first Zumba class and I loved it. It was perfection – the dance and fitness.

S: And you've told me in past conversations that you began training to be a vet. Tell me a little bit about that.

C: In Chile, I finished year twelve, and then went to university for two years to be a vet. From when I was little, that was my dream, I wanted to be a vet. I very much enjoyed the study, however, then I had to stop because our house got burned down in a fire, and Dad lost his job. So it became impossible to pay for it anymore. And Mum was here (in Australia), so she said to me, "You know what, come to Australia, give it a go." And I applied and three months later I got the permanent resident visa.

S: Fantastic.

C: Yes, it was meant to be.

S: You mentioned that it was in 2009 that you discovered Zumba. Did you know straight away that this was something that you might evolve into becoming a full-on passion?

C: In the beginning, I loved it because I was very unfit, and it's a great workout. And then I was going three or four times a week – I got a bit obsessed with it. Then my friend, she was an instructor, she said to me, "You should teach." And I said, "Really?" She said, "Sometimes you remember the choreographies better than me. Come on, give it a go and you can work for me." So I decided I would give it a try, and I remember I was very nervous. I did the training, I was very familiar the dances and moves, so the training was very easy. And then my friend organized my first class, it was at the bowling club at Mortdale, with only five or six people. I was very nervous before the class. Even now, I still get nervous before a class.

S: Interesting, so you still have a little bit of anticipation before you teach.

C: Yes. However, your confidence grows each time you teach. In the beginning, I was very nervous. But that first time went well.

S: We're talking about finding your bliss. What do you think the benefits were to you of following your dreams? You obviously had two paths you could follow – you could have carried on becoming a vet in Australia if you'd wanted to, but you chose Zumba. How has that impacted you, your health, your mental wellbeing, your children...how do you feel like it's been different, or better for you choosing that path?

C: It was a great choice, because at that time Stephanie [her daughter] was only one and a half, and I was locked in at home with the baby, and I was wishing for something for me. I wanted to do something for myself. And it was just perfect timing; I could earn a little bit of money and keep myself fit, because I had been getting quite unfit. My friend asked me to join her class, and then she asked me to work for her; it was awesome. I feel like I've found my passion.

Even when I was little, I was good at theatre. I was always in plays, and I was the first one singing, I was always in the front. And then when I started studying, I didn't do that much, so I think Zumba it brings back all my...capacity. I had the ability to stand up at the front, I can talk to people, I can express myself, and I think I'm contagious with my energy. So yes, I think it's been awesome, I feel fit. I think, not even when I was fifteen I was this fit. I am confident. And, it's my time. During the day, I give and give and give, to my girls, to the house, to my husband. But then, teaching Zumba, it's my time and I love it.

S: Just a little bit of background for the readers here. I'm a student of Cynthia's, so I see the enthusiasm when she comes to class, it's fantastic, she's got so much energy, and it's contagious, it spreads through the room. You can see that she loves dancing and she loves teaching. This energy and enthusiasm are so essential when you're teaching.

C: Imagine some days, you wouldn't know, but I have a bad night – my kids may have woken up at two a.m., four a.m., and six a.m. Then I have a busy day, and then I go and teach. So sometimes I'm very tired, but I don't mind, I still want to do it.

S: I've found that the people that follow their passion tend to have a lot going on – it's all stuff they love, so they're not giving off that 'I'm so busy' vibe. You've got a family, you've got to look after them, and you've got your Zumba classes all over the place on the Sunshine Coast. How do you fit everything in? Do you have any strategies to help you deal with all the things you have going on?

C: Yes, well my classes are on Monday, Wednesday, Thursday, and Saturday. For the Monday and the Wednesday, my husband gets home at five forty-five p.m. and I leave at six p.m. On Thursdays, the gym has a crèche, so I can take Amelia [her daughter] with me. And Saturdays, there's a crèche in Caloundra. And, the other one in Buderim, my mother-in-law helps me. She stays with the girls for an hour and a half. So that's my support network – my husband, and my mother in law, and the crèche.

S: So you have a little support structure set up to help you with it all.

C: Yes, and I'm busy, but I know Mondays and Wednesdays I'll do all the cleaning and cooking and shopping. And sometimes I help with Stephanie's school – I go on excursions. But I know, at six o'clock I have Zumba.

S: It's your escape.

C: Yes, that's it. I'm going.

S: One of the things I'm including in my book is myth busting. A lot of people don't follow their dreams because they feel like they can't make money, or they don't have time, or it's just a bit frivolous. One of the things I think is important is the financial viability of it. Do you find that it's been financially viable for you to do this? I'm sure being a vet would have meant a lot more money, but also a lot more time, and you wouldn't have had so much time for your children. Are you making enough money from Zumba to make it worthwhile for you?

C: You know what, I would do it for free!

S: Really?

C: I love it so much!

S: So many people that are living their passion feel like that – they love what they do so much they would do it for free.

C: Yes, I'm getting the fitness, the music, the chance to listen to Spanish music. It's my time, I teach it, I love it. So it's a plus that I have that cash in my pocket. You know, I'm in a good situation, I'm lucky that my husband has a good job, so I can stay at home and look after my girls, I don't have to go to work. This cash that comes from Zumba is my extra money, my pocket money, I can do what I want with it. It's not that much, but it helps.

S: It just adds to what you already have.

C: Yes, and it's a big decision to choose to follow your dreams. I chose to be a Zumba instructor even though I had the chance to study, but I said that I'd rather do this because I love it.

S: It's interesting that you said you'd do it for free. I seem to be finding, with anyone who is following their dreams, that the money is secondary in their minds. I know this is a challenging way to think for most people, but it's such a fantastic attitude. I truly believe it's the approach you must take, and while it may take some time, it's worth it for how happy it makes you.

C: I save the money, sometimes I use it to go to Sydney to see my friends. I open the bank account and there's, say, six hundred dollars. And wow, that's for tickets and dinners. With that money, I am going to pay for my Dad's flights, he's coming from Chile in August, and I'm so happy I can help him to get here.

S: Looking back over your life is there anything that you would have done differently, and I guess kind of linked to that, is there any advice you'd give to others that would like to follow their dreams, is there anything that you think is important for people to have that you've noticed, that you've seen in yourself?

C: I think it's easy not to follow your dreams. Just stay where you are and don't make a change. We all have busy lives, we all have schedules and things to do, but it's good for yourself, it's good for your soul, to make time and just to follow your dreams, you know.

S: So even if you're not doing it as a job, find something that's your passion and live that life outside of work. Just this small focus will make you feel better.

C: Yes, it will make you feel better and allow you to keep going in everyday life. If you have time for yourself, to do something that you are passionate about, you are better the next day, you feel ready for it, and with more energy and you're happier. You give and give, and you work, or you study, or whatever you do, you're so busy. So you deserve to do something for yourself. Something that you love.

S: You're doing Zumba and you love it, and you're doing it for you. Yet, isn't it amazing how when you're living your passion, while it's for you, you're giving so much to other people too.

C: Yes, and I love it when I see those smiley faces looking at me. You know they're loving it, they're enjoying it. So it's good for them and it's good for me. And you can feel it from the atmosphere in the room. Everyone is feeling the same vibe – everyone is happy, in their happy place.

S: Just a little side note. To be honest, I'm not a very good dancer, I'm pretty uncoordinated, but I'm so impressed with how you can get everyone synchronized in class. Sometimes, I think, wow, we could all be in a music video, we're looking not too bad at all. This is totally due to you, you seem to have this real talent.

C: In the beginning, everyone is a little bit lost and confronted. But when you get used to the class, or get used to the steps and the music, you can hear when the changes are coming. The lady that works here [at the university where the interview is taking place] gave me some nice feedback. She said she likes how I build that trust feeling, that she feels even if she goes right or left, she feels she's never doing anything wrong. Because if you go to another class, if you don't do the right move, you feel judged or something. She said she liked the ambiance I create. She feels it is a good group of girls and she feels confident with me. So that was nice, that feedback.

S: Finally, are there any traits that you think are important to have to follow your dreams? Things like, personality traits, resilience. I know you said before, it's easier not to follow your dreams. I guess to follow your dreams...do you think it takes a little bit of courage?

C: Yes, courage, and determination and persistence. So if you want to do something, just do it. Just follow that desire, it will make you a better person.

S: Thank you. Thank you so much.

Key Learnings:

- Grab opportunities when they are presented to you, even if they seem scary or make you nervous.
- Doing the hard things is what will bring you success.
- Doing what you love gives you energy.
- It's important to set up a solid support structure to ensure you can do what you love and do it well.
- People who live their passion would do what they love for free – money is not the key driver.
- It's easier not to follow your dreams than to follow them. However, it's important to find your passion and do it – whether it's for work or for play, it will make you happier and healthier.
- When you follow your passion, it not only benefits you, it benefits other people.
- Some key traits of those living their passion are: courage, determination, and persistence.

You can find Cynthia at:

Cynthia Zumba Sunshine Coast Facebook

5

NINE SIMPLE QUESTIONS

Well, I certainly hope that you found those interviews as inspiring and useful as I did. One of my readers noted that the three traits that standout from all these people I've interviewed are their insight, determination, and trust. There are some very brave people out there, not focused on money or what people think of them, but focused on how they can make a real difference in the world, and they've gone out there and done it. Which means that you can too.

I also want to say here, please do not beat yourself up if you are not living your passion or if you don't feel like you're living on purpose. I know it isn't easy to identify your passion or purpose. And many of us have commitments, such as family obligations or mortgage payments, that limit our freedom to test and try things out.

Speaking from my own experience, there have been plenty of times when I've wondered whether giving up a successful career and spending my life savings on a career change to live on purpose has been the right thing to do. It's a very big call. You must do what's right for you, and only you know what that is. What I would say, however, is that if you are determined – and use the questions below to determine if you are really ready for change – then you will find a way to make it work. Whether it's staying in your current job and experimenting in the evenings or weekends, or doing a course in what interests you, or taking a sabbatical from work to travel for six months and see what appeals in that free time. There are plenty of options – it's all about experimenting, and figuring out what's right for you.

1. Is what you're working on or in now giving you energy or taking it away from you?

This is the first question you should ask yourself. If what you do for work does nothing but take energy from you, then it's definitely time to consider a career change. However, if you find that your current role still energizes you, then maybe it's not a career change you need, but rather some small changes in your current environment, or changes to your relationships or your behaviour.

2. How do you deal with challenges?

This is related to the first question. Perhaps it's not a career change that's needed – maybe it's just a matter of changing your mindset or changing the way you deal with challenges. Do you see challenges as opportunities to learn and grow? When a challenge presents itself do you start by looking for solutions? The way you deal with challenges is unlikely to change just because you've decided to make a career change. Make sure you're dealing with challenges the best way possible now – experiment a little, see how you can change your behavior around challenges. Consider this question before you think about a career change.

3. What obstacles will you face in changing careers?

One question you should ask yourself is, "Am I willing to start over?" This is a huge challenge for most of us – are you ready to start over? For me, I had to consider whether I was willing to go back to school for six years to become a psychologist. Personally, I believe it's been worth the effort.

4. Do you have the support you need?

Changing jobs or establishing a new business can be a challenging and time-consuming process. We all need support to help to keep ourselves on track in a new venture. How supportive are your friends and family of the change you are considering? It certainly helps to have a team of people we can count on to cheer us on.

At this early stage, it may be worth asking yourself, "Who can help me?" No matter what type of career change you're considering, someone will have been there before you and done the same thing, or something similar. Find them and ask them what advice they'd give you.

This also relates to the importance of finding a tribe. Not all of us are surrounded by the positive, supportive people we need to help us through a career change transition. However, these days there is a wealth of wonderful, supportive, online communities that can provide advice, guidance, and positive affirmation. Whatever you do, don't work in a vacuum.

5. How well do you know yourself?

Are you aware of your values? It's important that whatever goals you set match your values. If your value is learning and growing, and you decide to set up your own business, it's very likely your value of learning and growing will be fulfilled throughout your journey. The benefit of values matching in this case is that whether your business succeeds or fails, you are still gaining plenty in what you learn and the personal growth you achieve along the way. You'll enjoy the process and feel good about it because you know that learning and growth are important to you.

Do you take time for self-reflection now? If you're not doing it now, then how will you find the time to do it when you're establishing your own business or growing into a new area of expertise? Self-reflection is important to help you truly know your own strengths and weaknesses. Self-reflection also helps you to ensure that your desires are yours, that you're doing what's best for you, not what you think others expect or what society expects. So check in with yourself – are you self-reflecting? Do you really know what it is you want?

And, of course, this is easy to say but not necessarily easy for us all to integrate. It's important to love yourself, to trust yourself, and to have confidence in yourself. What can you do to build that self-compassion muscle? Self-reflection can help you build awareness of your thinking patterns – are you coaching yourself with your thoughts or putting yourself down? Start a gratitude journal – you can find details on this on my website. A gratitude journal assists in training you to focus on the positives rather than the negatives.

6. Do you have the traits you need?

Or can you develop these traits in yourself? Some of the traits mentioned in the interviews were courage, determination, persistence, flexibility, self-honesty, authenticity, connecting to your inner power, patience, stubbornness, intelligence, love of learning, being a good listener, creativity, enjoying challenges, doing what you love, and having fun. How many of those traits do you have? Are there any on that list that you'd like to work on? Also, consider what your skills are, and whether any of these are transferable to a new career. You may have a strong transferable skill set that allows you to easily transition into a new career. If not, what skills can you start building now, so that when you move into a new career you already have some of the skills you need?

7. Are you willing to take things slowly, to go the distance?

Are you able to go step by step? You need patience for this journey. Do you have what it takes? Like everything, career change is about timing. When is the best time to make a change? For example, if you want to change jobs but you know the job market is flat, then it might be best to wait, take the time to develop a new skill set, and then change when the economy is more robust.

8. What is motivating your desire for change?

Are you motivated by money, or do you have a bigger vision in mind? Again, I will use the quote that Bodhi provided: "When you have a purpose greater than you, it frees you up to be the best you can be. It's no longer about ego, you give yourself permission to be as successful as you possibly can because you know your purpose is greater than that. It empowers you."

Do you have realistic expectations? Make sure you do your research, know yourself, understand the market, and understand what is required to make the move. Ask yourself, "Will I be happier?"

9. Are you ready?

If an opportunity were to present itself to you right now, are you ready and willing to grab it? Are you willing to take a leap of faith, to let go of fear and worry? Are you ready to live a life on purpose? Are you ready for the challenges and the amazing joy and benefits that choice brings?

I hope the stories you've read and the questions listed above help you to determine whether you're ready for a career change. Ultimately, the final decision is up to you. You need to decide whether you're living the life you want right now, or whether you're waiting for some amazing future to emerge. I hope the information contained in this book provides the guidance you need to make the most courageous step – to do what is right for you right now.

VIP CLUB

Building a relationship with my readers is the best thing about writing. I occasionally send out newsletters with information on upcoming books, recommended books to read, life improvement tools, and great daily rituals.

If you sign up to the mailing list I'll send you my first book Simplify Your Life for FREE.

Click to join my VIP Club
http://bit.ly/sarahoflaherty

PLEASE LEAVE A REVIEW

Enjoy this book? You can make a big difference.

Reviews are the most powerful tools I have when it comes to getting attention for my books. Much as I'd like to, I don't have the financial muscle of a New York publisher. I can't take out full-page ads in the newspaper or put posters on the subway.

(Not yet, anyway.)

But I do have something much more powerful and effective than that, and it's something those publishers would kill to get their hands on.

A committed and loyal bunch of readers.

Honest reviews of my books help bring them to the attention of other readers.

If you've enjoyed this book I would be very grateful if you could spend just five minutes leaving a review (it can be as short as you like). You can jump right to the page by clicking below.

Amazon - http://bit.ly/rfccamazon
All other distributors use link below
https://www.books2read.com/u/3JKwEE

Thank you very much.

ABOUT THE AUTHOR

Sarah has extensive experience in leadership, people management, mediation, and working with change. She is also the author of *Simplify Your Life*. Sarah's online home is www.sarahoflaherty.com. You can connect with Sarah on Twitter at @sarahof, and on Facebook at www.facebook.com/sarahoflahertymind, and you can send her an email at sarah@sarahoflaherty.com if you ever feel so inclined.

MORE BOOKS FROM SARAH

Simplify Your Life

Do you wish life was a little easier? Discover the secrets to a simpler, more satisfying life.

Is your life lacking purpose? Are you often stressed and overwhelmed? If so, then it's time for a crash course in the skills that will lead to a more meaningful life. Let successful businesswoman, coach, and author Sarah O'Flaherty be your guide.

Developed from the integration of hundreds of books, a multitude of personal development training formats, and a twenty-year career motivating people, Sarah has created a simple, yet effective, four-part process that will provide you with the skills and confidence you'll need for a happier life.

Each section is presented in a simple style, with tips and easy-to-adopt strategies that will teach you how to unlock your potential. And the best part is, you'll enjoy reading it!

Inside Simplify Your Life you'll discover:
   * How to identify your values, strengths, and passions for greater self-awareness and increased life satisfaction.
   * How to develop strong healthy relationships so you can benefit from your interactions.

* How to find your purpose or calling for a more meaningful life.

* How to un-complicate your life with some essential tools such as mindfulness.

* And much, much more!

Simplify Your Life is packed with straightforward, honest, and practical advice. If you enjoy easy reads that really add value to your life, then you'll love this book. Sarah takes you straight to the foundational aspects of life that, if you get right, will ensure a satisfying and meaningful life.

Unlock your true potential with Sarah's easy-to-follow guide today!

You can find Simplify Your Life here:
https://www.books2read.com/u/bxo6Pb

Fresh Start: A Guide To Eliminating Unhealthy Stress

Feeling overwhelmed? Do you have trouble sleeping? Are you feeling increasingly depressed? Could you be suffering from burnout? We all experience stress, but sometimes stress can leave us physically and emotionally drained. It can be difficult to regain a sense of balance in our lives.

Inside *Fresh Start – A Guide to Eliminating Unhealthy Stress* you'll discover:

- What stress is, so that you can understand whether you're affected.
- The difference between stress and burnout, so that you know which of these you're dealing with.
- The many sources of stress, the key triggers, and how to halt stress in its tracks.
- Different coping strategies, so that you can see how your current coping strategies might be modified for better results
- In the moment stress reduction strategies, so that you can lower your stress levels today.
- And much, much more!

There are answers. Discover how to manage unhealthy stress and start feeling more calm and peaceful. Let Sarah O'Flaherty guide you to a healthier, happier life.

Sarah O'Flaherty assists people in improving their job/life satisfaction, and working through career transitions. Currently training as a clinical psychologist, Sarah leverages the latest research and techniques for managing unhealthy stress to help her clients emerge from the chaos of stress and find balance and greater peace in their lives. In *Fresh Start*, Sarah teaches you about stress in any easy to understand format, with the hope of releasing you from damaging stress once and for all. *Fresh Start* will help you to transition away from stress, while maintaining your relationships, your job, your home, and your sanity.

Fresh Start is packed with straightforward, honest, and practical advice that can be your wake-up call to a new start in life. If you like easy reads that tell it to you straight, then you'll love having Sarah on your team.

Buy *Fresh Start* to help you return to calm and balanced living!

You can find Fresh Start by clicking the link below:

http://bit.ly/fsamazonus

DEDICATION

Many thanks to all those who assisted me in this process, including:

Mark Dawson and team, Lynn Ryan, Serena Clarke, Stuart Bache, Ramona Alba.

Start by doing what's necessary, then do what's possible, and suddenly you are doing the impossible.

—Francis of Assisi

Printed in Great Britain
by Amazon

16746449R00078